Liverpool Cowkeepers

Enjoy!

DAVE JOY

Dve Joy.

AMBERLEY

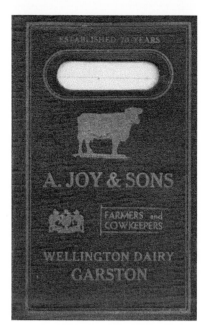

A Joy & Sons, Wellington Dairy. Account Book. (Author)

First published 2016

Amberley Publishing
The Hill, Stroud,
Gloucestershire, GL5 4EP

www.amberley-books.com

ISBN 978 1 4456 6322 7 (print)
ISBN 978 1 4456 6323 4 (ebook)

British Library Cataloguing in Publication Data.
A catalogue record for this book is available from the British Library.

Typeset in 10pt on 13pt Celeste.
Typesetting by Amberley Publishing.
Printed in the UK.

Contents

Preface

Garston, Liverpool, is where I was born and bred. When I was young my dad would tell me about three Joy brothers who, in the mid-1800s, relocated from the beautiful Yorkshire Dales to the expanding port that was Liverpool. Like so many Dales families, the brothers came to Liverpool not so much to make their fortune, but rather to make a better living – as cowkeepers, selling milk to the city's growing population.

One of those brothers was Daniel Joy, a widower with five young children – four girls and a boy. After the tragic death of his wife in 1872, Daniel decided to follow in the footsteps of his brothers, George and Orlando, and make a new life for himself and his young family. However, whereas his two brothers had set up business in the Liverpool district of Wavertree, Daniel relocated to Garston on the banks of the River Mersey. At that time Garston was a township and a port in its own right, separate from and lying to the south of the city of Liverpool.

Daniel Joy was my great-great-grandfather. His son was Anthony Joy and he would go on to establish a dairy in Wellington Street, Garston. As a boy, Wellington Dairy was a very special place for me. So much so that as a grown man I wanted to know more about how it came to be and about the lives of my antecedents who, riding the wave of the Industrial Revolution, exchanged the River Wharfe for the River Mersey. This is their story and the story of the many others who became Liverpool cowkeepers.

Anthony David Joy,
2016.

Wharfedale Roots

It has long been known within the family that the Joys had their ancestral roots buried deep in and around the Yorkshire village of Hebden, a charming Dales township nestling in the beautiful Upper Wharfe Valley. David Joy's delightful book *Uphill to Paradise* (1991) lovingly tells the story of Hole Bottom hamlet, Hebden. It includes a description of the coming together of two families – the Joys and the Bowdins.

The Joys of Rams Close

Rams Close was a farm located at the far northern end of the township of Hartlington, about a mile north of the Pateley Bridge–Grassington road. It lay in the Dibb valley, some two miles east of Hebden village. Part of the valley now lies beneath the Grimwith Reservoir.

The remains of New Lathe, Rams Close, Grimwith Reservoir in 2014. (Author)

The Joy family had farmed at Rams Close for many generations; the family can be traced back in the parish registers to the birth, on 7 February 1600, of Peter, son of William Joye (the 'e' had been dropped by the end of the seventeenth century). They were reputedly Calvinists who came over from France as a result of religious persecution. The Wharfedale and Craven Genealogical Study (http://wharfegen.org.uk) records several generations of Joys living at Rams Close, Hartlington, in the civil parish of Burnsall.

The Joys of Rams Close

Head of Household	Children
Christopher Joy (1672–1755) Marriage: (5 May 1892) to Grace Ackroyd (1667–1743) at St. Michael's Church, Linton.	John Joy (b. 1698) Elizabeth (1715–1716)
John Joy (b. 1698) Marriage: (7 November 1723) to Mary Knowles (1698/1708–1772) at St Michael's Chapel, Hubberholme, Arncliffe.	Thomas (1724–1767) Elizabeth (1725–1789) Grace (1727–1772) Mary (1729–1729/30) John (1729–1729/30) Anthony (1731–1801) Ann (b. 1734) Margaret (1736–1746) Christopher (b. 1741) William (1746–1746) John (b. 1748)
Anthony (1731–1801) Marriage: (24 August 1762) to Elizabeth Walker (1741–1820) at St Wilfrid's Church, Burnsall.	Anthony (1762–1830)

By the beginning of the nineteenth century the head of the household was Anthony Joy (son of Anthony and Elizabeth), who, in April 1796, married Nancy Hardacre, then of Hole Bottom Farm. Anthony and Nancy had seven children.

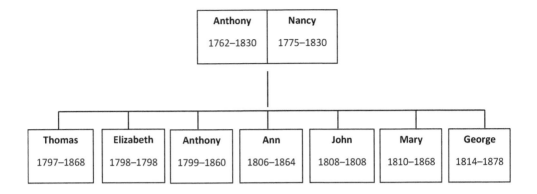

The Bowdins of Hole Bottom

It was in August 1796 that Thomas Bowdin and his wife, Elizabeth, moved to Hole Bottom Farm from their home in Ecton, Staffordshire. The property had been bought by Thomas's employer, the fifth Duke of Devonshire, who was investing in his lead mines on the moors above Hebden. Thomas had worked for the Duke at his mine in Ecton but had recently been appointed as the Duke's mineral agent for Grassington Moor. It would be his job to oversee the deepening of the lead mines and as such Hole Bottom Farm provided an ideal home for him and his family. Over the next seventeen years Thomas and Elizabeth had ten children:

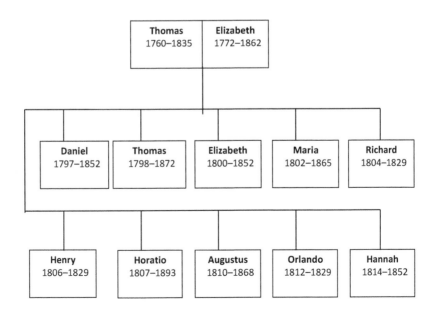

Marriages between the Joy and Bowdin families

Date/Place of Marriage	Joy Family Member	Bowdin Family Member	Children
4 March 1820 St Michael's Church, Linton.	Thomas	Elizabeth	Anthony (1820–1849) Thomas (1822–1893) Elizabeth (1823–1824) Elizabeth (1824–1891) Daniel* (1825–1896) George (1826–1904) Mary Ann (1830–1879) Orlando John (1833–1881) Horatio (1835–1836) Hannah (1836–1868) Unnamed infant (1838–1838)
20 May 1824 St Michael's Church, Linton.	Anthony	Maria	Mary (b. 1825) David (1826–1915) Anthony (1828–1883) Richard (1832–1908) Hannah Maria (1834–1899) Elizabeth Ann (1837–1905) Thomas Augustus (1841–1872) Margaret (1846–1918)

27 May 1828	Ann	Thomas	John (b. 1828)
			Anthony (1830–1895)
St Wilfrid's			Elizabeth (b. 1831)
Church, Burnsall.			Ann Selina (1833–1917)
			Daniel (1835–1920)
			Emanuel (1838–1912)
			Hannah (1840–1840)
			Hannah Maria (1842–1908)
			Mary Agnes (b. 1844)
25 May 1836	George**	Hannah	Elizabeth (1837–1894)
			John (1839–1904)
			George (1841–1865)
			Horatio (1843–1907)
			Anthony (1848–1921)
			Thomas (b. 1852)
No marriage	Mary	Horatio	Sabina (1833–1915)
			Horatio (1835–1884)

*The author's great-great grandfather

** Following the death of his wife, Hannah, in 1852, George married Ann Walker (widow of Ralph Newbold) and had five more children: Annie (b. 1865), Polly (1866–1884), George (1869–1891), Jerimiah (1872–1920) and Juliana (1874–1956).

Love is in the air

By pure coincidence, the Joy and Bowdin couples had their children around the same time span. And, although Hole Bottom Farm and Rams Close were separated from each other by nearly two miles of boggy moor, this was not enough to deter the respective offspring or to prevent young love from blossoming. So much so that, between them, the two families had four marriages and one common-law relationship, producing thirty-six children in total. As David Joy so eloquently puts it, 'this extraordinary liaison between two families living in virtual isolation seems to have everything needed to form the centrepiece of some great English novel.'

So many names from such a long, long time ago, yet, thanks to a family heirloom, it is possible to put faces to a few of these. The photograph (below) has been passed down the generations. On the reverse is written the names of those who were captured by the camera in their Sunday best, outside the farmhouse at Hole Bottom Farm.

Hole Bottom Farm, *c.* 1880. Anthony Joy, his mother, Uncle David, his wife, his sister, Uncle Dick and Horatio Joy. (Author)

Yorkshire Farmers

The Yorkshire Dales is an upland area of the Pennines. The typical Dales landscape is very picturesque, with rolling hills, deep-cut valleys, waterfalls and diverse wildlife. In contrast with the heather moorland that tops many of the hills, the valley pastures are verdant, with fields separated by drystone walls and dotted with farm buildings, all constructed from local stone. Words regularly used to describe the scenery, and quite rightly so, include 'dramatic', 'outstanding', 'magnificent' and 'breathtaking'. However, although the Dales may be easy on the eye, historically the climate, geography and geology of the area combined to ensure that making a living there was anything but easy.

Dales geology is predominantly carboniferous limestone, capped on the hills by millstone grit. The limestone is soluble in water, especially slightly acid rainwater, and so erodes without adding anything physical to the soil. Consequently, limestone soils are typically thin and lacking in nutrients. In some places the soil is so thin that the bedrock appears at the surface, creating scars or limestone pavements. Although it is a hard, non-porous rock, it *is* permeable – rainwater being able to flow through the rock's joints and planes. A consequence of this is that the soil on the valley sides being typically dry. Where surface run-off does occur, it will carry material down the steep hillsides and deposit it in the valley bottoms, giving a slightly thicker and more fertile soil there.

One of nature's quirks is that the lower the fertility of the soil, the more diverse the flora it can support. This is down to the life strategies of different plants, whereby a few nutrient-grabbing species are able to outcompete all others on rich, fertile soil but are unable to survive on poor soil – leaving the field to the many also-rans. Nutrient-poor soil produces nutrient-poor vegetation. So, there would be no growing crops in quantity in that situation. Indeed, it takes a particularly undiscerning and tough digestive system to make a meal from what grows on the hill areas of the Dales. One such digestive system occurs in sheep – hence their predominance in hill farming.

Down in the valleys, it is still not possible to grow crops on any significant scale, but the slightly deeper and more fertile soil can support a reasonable sward of grass and herbs. In his book *Britain's Green Mantle* (1949), A. G. Tansley distinguishes between pasture and meadow. Whereas the grasses of meadow are typically taller-growing species and are traditionally cut for hay, those of pasture are typically low-growing species that constantly form new shoots when grazed; traditionally grazed limestone grassland is often referred to as 'natural pasture'.

Both sheep and cattle graze these valley pastures; the unique digestive system of ruminants enables them to extract as much nutrient as possible from this plant material and convert it into meat and milk. Thus, although the landscape of the Dales is an agricultural landscape, it is one of pastoral rather than arable farming.

The climate and altitude of the area combine to present further challenges for the pastoral farmer. The year-round, higher than average rainfall can impact on haymaking in the summer and, at these altitudes, will fall as snow in the winter. Indeed, the growing season in the Dales is much shorter than in many other parts of England.

Despite the thin, nutrient-poor limestone soils, the rain-soaked summers and the long, cold Pennine winters, farming families in the eighteenth and nineteenth centuries still managed to subsist. They did so by hard work, applied by the whole family to both the management of the fields and the running of the farmhouse. In their book *Life and Tradition in the Yorkshire Dales* (1997), Marie Hartley and Joan Ingilby provide detailed descriptions of traditional daily life in the Dales.

Most of the sites of Dales farmhouses are ancient, their location being determined by the proximity to a supply of fresh water. The farmhouses at Rams Close were located close to the River Dibb. Originally there were two properties, known as Near Rams Close and Far Rams Close respectively. Today, all that remains of the farmstead is an old barn, 'New Lathe', which would have been located equidistant between the two properties.

In 1977 Far Rams Close farmhouse was surveyed by the North Yorkshire and Cleveland Vernacular Buildings Study Group. The group's report records and describes the building and also gives an indication of how it was once used by its early incumbents. The property, which has been dated back to the late 1700s, consisted of a traditional long house, in which the house and the cowshed were under one roof. The original long house and all subsequent extensions to it were constructed of well-dressed millstone grit. The design is described as being 'unusual' and 'eccentric'. As originally built, the house, located in the centre of the structure, had a living room on the ground floor, heated by a stack on the west gable end. To the rear were a kitchen (later extended, probably to function as a dairy) and, under the staircase, a larder. To the front was a separate parlour, probably for entertaining guests. From the living room, the centrally located staircase led to bedrooms above. The living room was accessed directly from the front door. To shield it, a single-storey porch with a lean-to roof had been constructed.

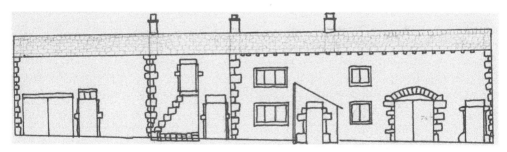

Far Rams Close – front elevation. (YVBSG)

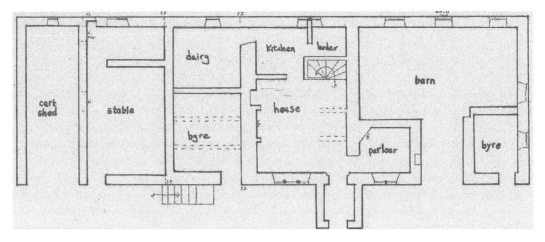

Far Rams Close – plan. (YVBSG)

To the east of the house there was a cowshed with eight stands and also a small byre, which could have been used as a milking parlour. To the west, the house was extended to give two ground-floor rooms, neither of which seem to be for accommodation. One was an extension to the kitchen and became the dairy, but the other seems to have been built to house equipment or possibly stock of some description (e.g., calves, pigs or poultry). Also, a bedroom was added above, reached by an external staircase. A further extension then provided a stable, with four brick stalls, and finally, a cart shed. All in all, it was quite a busy and versatile building. The occupants seem to have been industrious in making the most of what was available.

The house was the hub of all that happened on the farm. And within the house the centrepiece was usually the fireplace. At Rams Close the fireplace was located on what was originally the west gable end. It not only provided the principal means of heating the building but also the means of cooking food and boiling water. Where coal and timber were scarce or expensive, the main fuel for the fire would have been peat. Across the wide chimney was an iron reckan bar, from which were hung pots at different heights, according to whether a quick boil or slow simmer was required. A kail pot could be stood on the hearth and used for both boiling and as an oven for baking bread, pies, cakes and the ubiquitous oatcake.

The kitchen, larder and dairy were all located on the cooler north side of the building. The larder would have provided a cool room for the storage of raw or preserved foods. The kitchen would have been used for preparing raw food and mixing ingredients prior to cooking and also for preparing cooked or preserved food for mealtimes. Raw food for the kitchen came in many forms: milk from cows; eggs from chickens, geese or ducks; or meat was available whenever an animal was slaughtered -- beef, lamb, pork or poultry. After a slaughtering, nothing from the carcass was wasted. To preserve it, meat was cured by being pickled, salted, smoked or dried. It was common for cured sides of meat to be hung from the ceiling near the fireplace.

Most farmhouses had a garden where they grew fresh vegetables and oats. The fertility of the soil in the garden was maintained by adding profuse quantities of manure, sourced from both the shippon and stables and from the family's one privy. The kitchen would also have been the place for making cheese and butter, but at Rams Close the building was extended to include a dairy separate from the kitchen.

The creation of this dairy suggests that the farm was increasing its production of dairy food, especially butter, which needed space for the milk to stand while waiting for the cream to rise. Once the cream was skimmed off it was churned to separate the butterfat from the buttermilk. Churning also required space to work. It was done manually using either a plunge churn, a barrel churn or, more latterly, an end-over-end churn. After churning the butter was clashed and washed by hand to remove the buttermilk and then salted, weighed and shaped.

Although the process of cheese making did not require as much space as butter making, it did require space for the end product to be stored, preferably on shelves. Perhaps the larder at Rams Close acted as the cheese room. Both cheese and butter were sold at market or to agents who visited the farms.

Work in the house was synchronised with the work that was going on out in the fields. Most Dales farms had mixed stock, with cows in the pastures, sheep on the open moors and pigs in the sty – not forgetting the very important role that horsepower played in both the running of the farm and in providing the main means of transport away from the farm. The busiest times of year were lambing in spring and haymaking in late summer. Both were anxious times and success depended very much on the weather.

Far Rams Close – architect's drawing of former building superimposed on modern landscape. (Author/YVBSG)

A feature of cowkeeping in the Dales was that cows were kept indoors during the winter due to the severity of the weather. They were fed on hay and milked twice a day. 'Hay' cheese, made in winter, was known to be rather tough, so many farms put their efforts into butter making while the cows were indoors. Cheese making would resume in May when the cows were returned to the pastures. Ideally, cows would calve in the spring and produce lots of milk throughout the summer.

During the spring and summer, the twice-daily milking would take place in the field and milk would be carried back to the farmhouse in a *backcan* – a tin can, specially shaped for carrying on the back by both men and women. Cows would remain grazing outdoors in the pastures until October. Clearly, if a milk herd was to survive, it had to have enough hay to feed it through the winter.

While some fields were kept as pasture for the cows to graze during summer, others were kept as meadows. Stock was excluded from the meadows from May onwards in order to allow the grass to grow. The low fertility of the soil meant that the meadows were herb rich, with a great variety of wild flowers growing alongside the grasses. That might bode well for the naturalist but, for the farmer making hay, these herbs, being more succulent than the grasses, took longer to dry out and so prolonged the process.

Depending on the weather, the meadows would be cut in July or August – dozens of men working with scythes. Once it was cut, it was left lying to dry. To quicken the drying process all the hay had to be turned and tossed or 'fluffed'. Finally, when properly dry, the hay was taken to a barn for storage over winter. The whole process needed a good four to five days of rainless sunshine. If it did rain, then drying had to begin all over again. The time constraints made haymaking a very labour-intensive operation, with all members of the family involved along with hired help. It was common practice for those who finished first to go and help out a neighbour who, on account of the altitude of their farm, may have had a late harvest. It was essential for survival that everyone was able to bring in enough hay to see them through the winter. In this way haymaking was very much a social event among the farming community and its completion was a time for celebration.

The Dibb Valley,
Wharfedale in 2014.
(Author)

The Dales Exodus

Thomas Joy and Elizabeth Bowdin were married on 4 May 1820 at St Michael's Church, Linton. Over the next eighteen years they had eleven children:

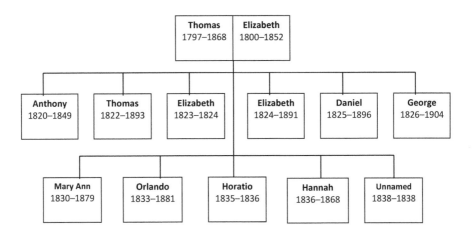

Initially, Thomas and Elizabeth lived at Hole Bottom Farm until after the birth of their first child, Anthony, in October 1820. They then moved on to farm at Long Lands in Hebden village before moving on again to Dibbles Bridge, located just outside Hebden where the road to Pateley Bridge crosses the River Dibb – near Rams Close. The Burnsall Parish Register records Thomas and Elizabeth's seventh child, Mary Ann, as being christened on 15 August 1830 and the family as still farming at Dibbles Bridge. However, by the time of the christening of their eighth child, Orlando John, on 5 May 1833, Thomas and his family were living back in Hebden village and he had become an innkeeper. He was the keeper of the inn then known as the Clarendon, which at that time was owned by Daniel Bowdin – Thomas's brother-in-law.

Despite his status as innkeeper, Thomas continued to be a farmer. Indeed, in order to maximise their earning potential, it was common for people to have more than one

Hebden High Street in 2014. (Author)

Former Clarendon Inn, Hebden, in 2014. (Author)

occupation. For, although the setting may have been idyllic, life in the countryside most certainly was not. It meant poverty, subsistence wages, and backbreaking manual labour shared by men, women and children working long hours.

Thanks to the deposits of lead ore under Grassington Moor, Hebden, like many villages in the Dales, had something of a mixed economy from which the local population benefited both directly and indirectly. The direct benefit was the opportunity of employment in the mining industry. Many of the village's menfolk, who more than likely would otherwise have continued in their family tradition of farming, were able to gain employment in the mines, albeit at poor levels of pay. The indirect benefit was the opportunity to provide goods and services to the influx of mine workers from outside the village. As well as providing a ready market for locally farmed produce, the newly swollen population also needed shops and a good inn or two, providing employment and business opportunities for local men and women alike.

As pay was so poor it was not surprising that some families kept more than one iron in the fire, to which the census records bear testament, with 'occupation' being described as 'farmer and miner' or 'farmer and innkeeper' or some other combination. Others witnessed a generational split, with the sons and grandsons of farmers going down the mine on a full-time basis. These changing trends in employment can be seen in the Joy family tree.

Anthony and William Joy were brothers who, as boys, had been trained in the craft of lead mining by their father, Christopher, himself a lead miner. Underground, they worked a near vertical shaft or 'rise', driven upwards from the main level to follow the vein of lead. As it was the Joy brothers who worked this particular shaft, it became known as 'Joy's Rise'. The work was long, hard, dark, dirty and dangerous, and was poorly paid. Perhaps it is not surprising that Daniel, George and Orlando chose a different path.

By the time he was in his twenties George was no longer living with his family in Hebden. The census of 1851 records the twenty-four-year-old George working as a general servant to Thomas Frost, Milkman, of 29 Peel Lane, Cheetham, Manchester. It was perhaps his experience of working a milk round in Manchester that sowed the seeds in George's mind of a way of making a better living than could be had from either farming or mining. And George was not alone in his thinking.

Those Who left the Dales (2010), by The Upper Dales Family History Group, is a collection of family histories gathered together by the descendants of those who made the great decision to leave their homes and seek a new life for themselves. These families were not solely from Yorkshire. In her unpublished thesis, *Origins of the Liverpool Cowkeepers* (1982), Joan Grundy uses the term 'the Dales' to include the following: Westmoreland; the West Riding Registration districts of Settle, Skipton and Sedbergh; and, Lancashire north of the Lune, including east Lancashire from Pendle Forest northwards. Throughout that area, the nineteenth century was a period of depression, due in part to an agricultural slow down, the Enclosure Act (which made the rich richer and the poor poorer) and the gradual closure of the lead mines.

In Hebden the ore veins were beginning to dwindle and the mining could no longer provide a living for the number of miners it had supported in its heyday. Miners began to leave the village in search of employment elsewhere and, as they did so, they took their

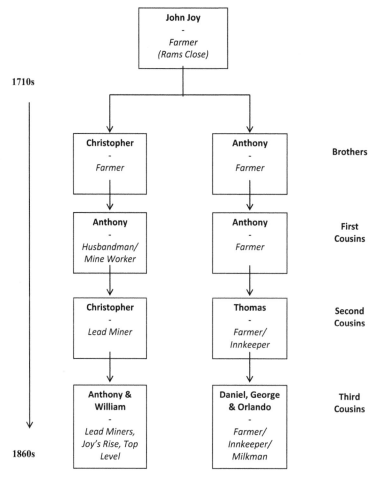

1710s

John Joy - *Farmer* *(Rams Close)*		
Christopher - *Farmer*	**Anthony** - *Farmer*	**Brothers**
Anthony - *Husbandman/* *Mine Worker*	**Anthony** - *Farmer*	**First** **Cousins**
Christopher - *Lead Miner*	**Thomas** - *Farmer/* *Innkeeper*	**Second** **Cousins**
Anthony & **William** - *Lead Miners,* *Joy's Rise, Top* *Level*	**Daniel, George** **& Orlando** - *Farmer/* *Innkeeper/* *Milkman*	**Third** **Cousins**

1860s

Changing trends of employment in the Joy family.

spending power with them. Consequently, those who had made a living from selling goods and services to the mining community began to suffer further hardship.

Of course hardship was nothing new to those families who had farmed in Wharfedale for generations. They had always had to eke out a living from the thin soils of the Dales. Nevertheless, for them the effects of the depression were compounded by having to house ever-growing extended families in comparatively small farmhouses, with not enough food to go around. Some resolved to cling stoically to their agricultural roots, but for others the thought of living a life in continuous hardship with little prospect for improvement led them to begin to reassess their situation.

In contrast to what was happening in the countryside, the cities were booming. In the mid-1800s the Industrial Revolution was proceeding apace. New industries were emerging in the cities and this in turn was pulling in more people from the surrounding countryside in search of work.

Clearly, if George Joy were considering setting up in business as a milkman, he would not be doing so in Hebden; he needed to be where the people were – in the city.

Farmer Turns Cowkeeper

The city of George's choice was not to be Manchester, where he had learned his trade, but rather the rapidly expanding port of Liverpool. At that time Liverpool was building its economy on being one of the busiest seaports outside of London. This situation is captured in the introduction to *Pigot's Directory of Liverpool* (1828):

> Liverpool carries on a trade to most parts of the world, and is a depot for East and West Indian and American produce. From this port the various manufactures of Manchester, Birmingham, Sheffield, Staffordshire, Leeds &c. are spread over the surface of the habitable globe. By inland navigation Liverpool communicates with every principal town in the kingdom, and by this means receives additional wealth and importance. The manufactures of Liverpool are principally refined sugars, glass, watches, soap &c. there are also salt works, copperas works, iron works, and many considerable breweries. Near the town are many windmills, for grinding corn and other articles. An immense number of shipwrights or carpenters, rope makers, sail makers, &c. are also constantly employed in making, repairing, and equipping vessels. The rapid extension of commerce, with its attendants, wealth and population, constitute the great features of the town and port of Liverpool.

With its busy docks, flourishing seaport, new industries and a recent influx of immigrants escaping the famine in Ireland, the growing population of Liverpool created a huge demand for food, especially fresh food and, in particular, for fresh milk. In an article for the *Sedbergh Historian* published in 2002, John Holmes describes how local farmers could not keep pace with the demand and how milk transported by rail from outside the area had problems with souring. The simple answer was to keep cows in the city, producing fresh milk at the point of delivery, but it seemed that the local city populations lacked the cattle management skills and know-how to do this. There was a ripe business niche here and the farmers from the Dales had just the right skill set with which to exploit it. Consequently, as part of the mass exodus from the countryside, many farmers made the journey from the Dales to become cowkeepers in Liverpool.

These farmers made this journey by rail. Farmers in Wharfedale were able to access the rail network at Skipton. Skipton station was opened in 1847 and in the following year became connected to Colne via the Midlands Railway. Subsequently, in 1849, Liverpool became accessible from Colne when the latter was connected to the Lancashire & Yorkshire Railway (Marshall, 1972). Other parts of the Dales became linked via the railway network when the Skipton–Settle–Carlisle Railway opened in 1875, traversing the region from north to south. The availability of relatively cheap travel by rail meant that Dales farmers could not only bring their whole family with them, but their cows as well.

This migration to Liverpool was not a straightforward relocation from one place to another; it was a 'chain migration', involving many family members doing a great deal of to-ing and fro-ing between the home farm and the newly founded city cowhouses. In her family history *A Bit Akin – The Story of a North Craven Farming Family* (1994), Faith Finegan describes the migration of her family to Liverpool:

> The first Wolfenden to be attracted to Liverpool was Cousin Robert ... He went to Bootle about 1877 with his wife, Jane, and their young baby. They would have plenty of family support. Two brothers-in-law, James and Robert Waterworth, were already there: they had married cousin Robert's sisters, Betsy and Isabella. Some of Jane's family, the Manserghs, were there too.

Once a foothold was established, other members of a family or extended family would soon follow:

> News soon spread that there was money to be made in Liverpool by those who were prepared to work hard and other Wolfendens went. Robert's elder brother, James, who had been farming at Phynis near Slaidburn, took his wife and large family to Liverpool in the 1890s where they kept a milkhouse in Waterloo... About the same time Tom Wolfenden left Studfordgill and joined his sons in Bootle, they had a milkhouse at 19 Brown Street, within sight of the docks. My grandfather's brother, Charlie, had left Orchaber, near Austwick, in the mid [eighteen] eighties and set up as a cowkeeper at 1 Beech Street in Bootle.

This migration was so successful that Dales folk began to form their own community within Liverpool. Initially, they were thought of as foreigners as their country drawl was so different from the dialect spoken by the native Scousers. In his memoir of a Liverpool childhood between the wars, *Candles, Carts & Carbolic* (2011), Jim Callaghan describes his cowkeeping neighbour: 'John the Cowkeeper was a Yorkshire man, a "yowk", my father called him, by which I suppose he meant a yokel...'

In another article for the *Sedbergh Historian*, dating to 2008, Joyce Scobie describes the basic physical requirements of a successful cowkeeping business. Cowhouses would have included the following: a shippon in which to keep and milk the cows; a dairy for butter and cheese-making; a shop, as a point of sale; a cellar for keeping the produce cold; a midden in which to keep the muck; a hayloft over the horse's stable; and a yard in which to keep a cart. The owners would probably live over the dairy. There is a great similarity between this combination of built resources and that which had already existed at Rams Close for over a century. It is alleged that, in Liverpool, these new cowhouses popped into existence on the end of just about every street of terraced houses, but perhaps that is an exaggeration.

Cowkeeping was by no means an easy way to make a living. It involved long hours, seven days per week, fifty-two weeks per year. But Dales folk were no strangers to that kind of arduous work – out of necessity, the Dales farmer was typically determined and self sufficient. First milking took place 'at the crack', usually 5.00 a.m. and by the time that was completed and the milk loaded onto the waiting horse and cart, first delivery could commence around 7.30 a.m. Second milking took place at 2.00 p.m. and second delivery commenced at 4.30 p.m. Prior to the invention of glass milk bottles, delivery was on the doorstep, the milk being dispensed into the customers' own jugs from the 'kit' (7-14 gallon churn), using either a measured ladle or a 2-pint or quart delivery can. Jim Callaghan (2011) recalls the activities of his neighbouring cowkeeper:

> The shippon, which housed his six cows, was directly under our back bedroom and looking down on them through the knothole in the floorboards we could feel the heat rising from their bodies ... As a child I would toddle round to John's dairy to get a warm drink straight from the cow and to watch him cooling the milk which he poured through a funnel into a steel ventilator, from which it emerged through the vanes directly into the churns ... When I became older I sometimes helped on his rounds. That was when housewives bought milk straight from the churn and a valued possession was a white quart jug.

To a degree there was a division of labour. Milking and delivering tended to be the domain of the menfolk, although not exclusively so. However, there was a lot more to cowkeeping than just milking and delivering. The women of the family ran the shop, selling milk direct to the customers in the immediate vicinity of the dairy. They would make and sell butter, cheese and cream and, if the premises were large enough to accommodate hens and a couple of pigs, they would also sell eggs and bacon.

Another important duty for the women was maintaining the cleanliness and hygiene of all aspects of the milk-producing operation. In his book *The British Milkman* (2011), Tom Phelps describes the poor reputation of milkmen in the early part of the nineteenth century. The levels of illness and death due to the consumption of infected or adulterated foods was so high that it led to the introduction of legislation and the enforcement of standards. Daily tasks included all work surfaces in the dairy being scrubbed, all utensils and vessels involved in milk production being scalded, and the floors of all rooms being washed before 9 a.m.. Additionally, the conscientious cow-wife would ensure that tiles were polished, steps were 'soapstoned' and all aprons and coats were washed and bleached. A further requirement was that all walls and ceilings were to be whitewashed twice per year. Any room for the preparation, storage or sale of dairy products could not be used for or directly connected to any living quarters. Premises that were granted licenses to sell milk had to display a notice over each entrance with the words: 'Registered For The Sale Of Milk'.

The regulations also applied to shippons. In her article, 'Purpose-Built Premises for Town Cowkeepers in Liverpool' (1990), Joan Grundy describes the shippon (where the cows were housed) as being the most important building on the premises. They were subject to their own regime of inspection and licensing. All buildings had to be available for inspection at any time and had to be kept clean, wholesome, properly drained and supplied with water. The walls of shippons had to be lime washed once per year, in March. Again, once a license was granted, a notice had to be displayed on the building stating: 'Licensed For Keeping x Cows'.

Traditional milking at Dugdale's Farm, Grassendale Road, Garston, in 1950. (Frank Smallpage)

Then, there was looking after the animals. The cows were mostly shorthorn, referred to as 'The Dual Purpose Cow' as they gave a good milk yield and would fatten up easily to be sold for beef. These animals had to be fed, watered and cleaned. Their (high protein) diet was a mixture of what was available locally: hay and grass from farmers on the edge of the city, made 'sweet' by adding molasses from the sugar refineries; grass cuttings from local parks, playing fields and sewage farms; bran from the flour mills; spent grain from the many breweries (in 1882 the city was home to fifty-six breweries and two distilleries, all producing spent grain); oilseed cake from the oil mills (e.g. linseed cake could be bought from the Kent Street oil mill, owned by Alexander M. Smith & Co.); Indian corn and peas imported in large quantities and at minimum cost from America and Canada; and vegetables from local farms and markets, including winter turnips from Cheshire and the Isle Of Man. The cows of Liverpool benefitted from an abundance of fresh, clean water, piped directly into the city – initially from Rivington Pike in Lancashire and later from Lake Vyrnwy in north Wales.

Much of this 'proven' was mixed with warm water and then allowed to stand while it swelled. It was then given to the cows to keep them still during milking. The art of the cowkeeper was to mix the ingredients in the appropriate proportions and serve them up in the appropriate quantities – this was the key to keeping healthy cows that produced flavoursome milk.

Wherever possible, the cows would be allowed to graze, at least during the summer months. In the densely populated inner city, this was not always practical. It was, however, possible for cowkeepers to come to mutually beneficial arrangements with local landowners whereby the landowner had their grass 'cut' and their land fertilized for free. There are records of cows being grazed in parks and also on land that would become Goodison Park. It was a constant dilemma for the cowkeeper whether to locate the cowhouse nearer to where most customers lived or to locate it a bit further away from the city centre where grazing might be available. Those that preferred a grazing location found themselves being pushed further away from the centre as the city expanded.

Mucking out involved shovelling up the cow muck and barrowing it out of the shippon into the midden, either ready for transportation to the Haymarket or for collection by the muck merchant. The muck was often sold on to local farmers or used in part exchange for oat and wheat straw, to be used for bedding. Sawdust was obtained from the local sawmills and used in the gutters and passages of the shippon to absorb water from the manure – thereby making it more manageable for transit. Staner (1882) provides an interesting equation to demonstrate the value of muck to the cowkeepers at that time:

- Four cows produce 1 ton of manure per week
- 5,016 cows produce 1,254 tons per week or 65,208 tons per year
- At an average price of 5 shillings per ton, annual income = £16,302

Common ailments among cows that were kept in confined spaces for prolonged periods were digestive problems and stiff joints. These symptoms were often remedied by haltering the cow and walking it around the neighbouring streets – much to the consternation of the local city dwellers.

There was also the horse, the business's main form of transport. Though shire horses were very popular with Liverpool carters for heavy haulage, the horse of choice for lighter transportation was the Irish Cob or 'Vanner'. The term 'Vanner' referred to a breed of horse

'Waiting' by Judy Boyt. A life-size bronze statue to commemorate Liverpool's working horses. Commissioned by Liverpool Ex-Carters Association. (Author)

Atkinson dairymen, Heathfield Road, Wavertree in 1915. (Mike Chitty, The Wavertree Society)

created specifically to pull gypsy caravans or tradesmen's wagons. Characteristically, they were strong and mobile and could maintain a steady, economical gait for hours at a time; they were intelligent and took instruction very easily with quick response; they could live on limited grazing and had a calm temperament; they were uniform in colour but with the occasional white face marking and had a good feather of hair on each leg. The horse would be turned out at its very best, its coat curried, dandied and towelled until it shone and its harness polished until you could see your face in it. The cart, or float, was painted and varnished and its livery proudly announced the name of the business for all to see.

A flourishing cowkeeping business was something in which the entire extended family participated, including those still living in the countryside. For many, the city cowhouse was operated as an extension or satellite of the main farm back home in the Dales. It was common for in-laws and cousins or nieces and nephews or aunts and uncles or brothers and sisters to come out to Liverpool to help with the family business before returning home. In a reciprocal arrangement the children of the Liverpool cowkeepers would spend their school holidays with their grandparents, back in the Dales. And family ties were not the only reason for cowkeepers to return home for a few days. They also travelled back to the Dales to do business, as the trainloads of cattle arriving at Lime Street station bore witness.

In her article 'Cow-keepers from the Pennine Dales' (published in the *Dalesman* magazine in May 1978), P. J. Mellor describes this buoyant trade in cattle:

Cattle for the milkhouses were bought, newly calved, often fourth, fifth or sixth calvers and kept throughout lactation. This varied but was usually about 18 months. Cows were usually replaced when their yield dropped below three gallons a day. They were then either sold for beef or sent to the country to re-calf. By the end of the nineteenthC 4,000 head of cattle were being kept within the Liverpool boundary and replacement stock was being bought from cattle marts as far away as Oswestry and Kendal. Cattle were driven through Liverpool by the dealers and dropped off, a few at a time, at the various milkhouses, to the consternation of the local population: "There's a bull loose" was a common cry.

'There's a bull loose!' Cattle being herded along Lime Street, Liverpool city centre, in 1931. (British Pathe)

In his essay 'A Short Account of Lancashire Farming in 1899', Daine (1900) notes that the chiefly cross-bred, short horn cattle were being purchased in considerable numbers from Cumberland, Westmoreland, the Craven district of Yorkshire and also from Irish cattle dealers. He goes on to describe a good milking cow as follows:

Long in her sides, bright in her eyes,
Short in her leg, thick in her thighs,
Big in her ribs, wide in her pins,
Full in her bosom, small in her shins,
Long in her face, fine in her tail,
And never deficient in filling the pail.

In his article 'The Dual Purpose Cow at Liverpool', Mackenzie (1910) goes further and identifies two distinct varieties of shorthorn being kept in the city – one originating from the Dales and the other from Ireland:

Two distinct types are to be found. One of these, though finer in the chine and immensely superior in the 'bag,' is very similar to the 'show' animal of Coates' Herd Book. Mossy in coat, somewhat thick in the hide, heavy in flesh—sometimes even patchy to a certain extent—there is no mistaking the presence of Booth, Bates, and no doubt a little Cruikshank blood, in these animals which are very often a rich roan in colour. The other type, though distinctly shorthorn, is less so than the first one. Finer of skin, hair, and horn, more slender of frame, often red and white in colour, she is decidedly more dairy-like in character. We learnt that not infrequently cows of this description were obtained from Ireland.

As well as spending their hard-earned cash back in the Dales, the cowkeeping businesses also benefitted the city economy in that they generated their own local supply chains. Trades that grew through their association with the cowkeepers included: cattle dealers, feed merchants, local farmers, auctioneers, cartwrights, wheelwrights, saddlers, blacksmiths, farriers, muck merchants, butchers, vets and many, many more. The wily cowkeeper would keep the best muck – horse manure – to one side in order to sell by the bucketful, sackful or cartful to local gardeners and allotment holders.

These immigrant farmers found cowkeeping in the city to be a profitable business – certainly more so than farming back in the Dales. They had a ready market for everything that they could produce. As they ploughed their profits back into their business they were able to expand: increasing the number of cows being kept; increasing the spread and number of milk rounds; and, inevitably, increasing the number of dairies.

Cowkeeping became an attractive proposition for the next generation. Back in the Dales, the farming tradition was to set up each son of the family in their own farming business and this tradition transferred to the city cowhouse business. However, whereas the son in the Dales would have to wait to get married until his family could afford to set him up in business, cowkeeping was relatively easy to set up and the cowkeeper's son did not have to wait to get married.

Initially, the cowkeepers looked to adapt existing buildings. The most affordable and most easily convertible were end terraced properties. These had the advantage of the back yard being accessible from the side street – usually by knocking through the sidewall to the yard and installing a set of gates. Also, they often had enough land at the side to be able to extend and build the additional rooms required for a fully operating dairy and family home. By their nature, terraced properties were in the most densely populated parts of the city – this suited the cowkeepers well, as it meant more customers located close to the cowhouse.

Later, as the city expanded, town planners recognised the need to accommodate this trade and would provide purpose-built 'cowhouses' with house, dairy, shop, yard, shippon, midden and stables all being included in a single complex. Two of the best remaining examples of these in Liverpool are the former Harper's Dairy in Rose Lane, Mossley Hill, and the former Batty's Dairy on Aigburth Road, Aigburth.

Greenwood's Dairy, High Street, Wavertree, in 1887. (Mike Chitty, The Wavertree Society)

Purpose-built cowhouse –
Harper's Dairy, Rose Lane,
Mossley Hill, 2015. (Author)

Rear yard – Harper's Dairy,
Rose Lane, Mossley Hill, 2015.
(Author)

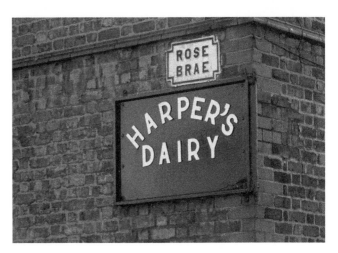

Business sign – Harper's Dairy,
Rose Lane, Mossley Hill, 2015.
(Author)

Membership of the National Dairymen's Association – Harper's Dairy, Rose Lane, Mossley Hill, 2015. (Author)

Purpose-built cowhouse – Batty's Dairy, Aigburth Road, Aigburth, 2016. (Author)

Rear yard – Batty's Dairy, Aigburth Road, Aigburth, 2016. (Author)

Doorway sculpture – Batty's Dairy, Aigburth Road, Aigburth, 2016. (Author)

The number of listings in the trade directories for Liverpool gives an indication of how the cowkeeping business was booming:

Kelly's 1894	818	'Cowkeepers' and 'Dairymen'
Gore's 1900	937	'Cowkeepers' and 'Milk Dealers'
Gore's 1911	1,039	'Cowkeepers' and 'Dairymen'

As the businesses expanded, more relatives from back home in the Dales would board a train and come out to do their bit and share in the bounty. A search of the 1881 census for those whose occupation is listed as cowkeeper, with their place of residence listed as Liverpool (including Garston), produces a list of 283 families.

It is said that Dales folk have a good nose for business and some were quick to exploit other business opportunities associated with cowkeeping. Some went on to specialise in trades such as cow or horse dealer, butcher, feed supplier, or muck merchant. As these businesses came into being, so even more family members were recruited to help in their running. In this way the Dales community presence increased and they became a key part of the Liverpool way of life, adding to the growing number of sub-cultures within the diversifying city.

Nineteenth-Century Milk Wars: Statutes, Shows and Silverware

The practice of keeping cows in a town or city was by no means limited to Liverpool; it was common in most built-up areas in the country during the 1800s. For example, in 1851 there were sixty cowkeepers in Hull and this number had increased to 260 by 1886 (Harris, 1977). However, once a town or city became connected to the rail network, its cowkeepers found themselves in direct competition with the companies that were supplying milk via the railways. And competition was fierce.

These companies operated by collecting milk from rural farms and transporting it in churns to railway stations where it was loaded onto trains bound for the cities. Once in the city, the milk churns were unloaded and distributed to dairies or 'milkhouses', from where it was sold to the public. In the absence of any form of refrigeration, this 'railway milk' was souring throughout its journey.

In Liverpool, railway milk was delivered on a daily basis to all of the city's main railway stations: Lime Street, Edge Hill, Exchange, Pier Head, Sandhills and Central. In 1882, there were 1,145 non-cowkeeping dairies in the city registered to sell railway milk. Frederick Stoner (1883) compares the supply of milk to the city from these two sources. At that time, railway milk accounted for some 37 per cent of all milk consumed by the people of Liverpool.

A major reason for the city cowhouses being so important in the milk supply of Liverpool is, probably, the maritime location of the city. An inland city such as Manchester

Comparison of milk supply from different sources – Liverpool, 1882		
Source	*Gallons per Day*	*Gallons per Year*
City cowhouses	15,048	5,492,520
Railways	8,792	3,209,080

has a hinterland of 360 degrees from which to draw its supplies of food, whereas a coastal city such as Liverpool only has 180 degrees of hinterland from which to draw its supplies. Consequently, with regards to meeting its demand for milk, the city had to rely on the supply from within its boundaries.

The occupational title 'cowkeeper' was not accidental. It was adopted deliberately in order to help differentiate the dairies that kept cows from the dairies that did not keep cows but supplied railway milk. The cowkeepers were producer-retailers who were in competition with suppliers of milk who were purely retailers. The cowkeepers' principal selling point was that their milk was fresh, straight from the cow, whereas railway milk could be up to two days old by the time it reached the customer and during that time had been constantly souring – especially so if it had been left standing in direct sunlight, waiting to be collected.

Towards the latter half of the nineteenth century there developed a nationwide concern about the extent of adulteration and contamination of the milk supply and its impact upon the health of the population. This concern grew into a national movement for 'clean milk' and an advocacy for a more scientific approach to management. Among the most vocal were the self-named 'progressive' dairy companies, such as the Aylesbury Dairy Company, which made extensive use of the rail network to transport its milk into London. However, the motives of such elite dairy companies may not have been solely about improving the quality of milk. Many commentators were of the opinion that the corporate dairies were using 'the tools and the rhetoric of scientific management to gain market hegemony' and, in so doing, were marginalising the smaller milk producers (Steere-Williams, 2015).

They were particularly critical of city cowkeepers in general, not only accusing them of adulterating the milk with (contaminated) water, but also of keeping poorly fed and unhealthy stock in unclean premises, and of using unhygienic practices for the extraction, storage and conveyance of milk. Whatever their motives, it seems they did have a point. In his book *Drinka Pinta – The Story of Milk & the Industry that Serves It* (1970), Alan Jenkins described the consequences of the general lack of understanding of the most elementary principles of hygiene that prevailed during the mid-1800s in London and elsewhere: milk was carried in uncovered pails through the dusty streets; cows were often kept in the back of milk shops; some were milked in the streets; many were diseased, their sheds filthy, the pails rusty and unscoured. There were also fearful tales, not only of watered milk, but of milk with fish actually swimming in it.

Between 1865 and 1867 the country experienced an outbreak of the cattle plague 'Rinderpest'. In London, the disease served to highlight the poor standard of city cowhouses and many of these were subsequently closed – this opened up the London market for the corporate dairies supplying railway milk. A different outcome was experienced in Liverpool. Firstly, the disease devastated the Cheshire herds, which at the time were the main source of railway milk supplying Manchester and Liverpool. Secondly, probably due to the work of successive medical officers for health, Liverpool's city herd was virtually disease free (only five cases recorded). This combination of circumstances served to strengthen the position of the Liverpool cowkeepers in their battle with the corporate dairies.

This rivalry between the two factions went on to become quite politicised, with either side arguing for or against legislation according to whether or not it was seen as being to their advantage/disadvantage or the advantage/disadvantage of the other side. Although this debate was destined to rumble on for the next hundred years, it began by yielding a number of Acts of Parliament and the introduction of orders and regulations designed to improve the health of the people. A summary of the main legislation affecting cowkeepers in the nineteenth century is included below.

Nineteenth Century Legislation Affecting the Liverpool Cowkeepers

Year	Legislation/Regulation	Impact
1860	Adulteration of Food & Drugs Act	Power to appoint public analysts.
1867	Liverpool Improvement Act	Licensing and inspection of premises where cattle were kept and milk sold. Obligation to safeguard the health of animals.
1872	Adulteration of Food, Drink and Drugs Act	Made the appointment of public analysts mandatory.
1875	Public Health Act	Public Health Act 1875 provided powers to inspect and seize unsound food, including milk.
1875	Sale of Food and Drugs Act	Confirmed offences of strict liability and introduced heavy penalties for adulteration of food including three months' hard labour for a second offence.
1878	Contagious Diseases (Animals) Act	Made it possible (not compulsory) for local authorities to register all cowkeepers, dairymen, and purveyors of milk, to regulate the lighting and ventilation of cowsheds, and to secure the cleanliness of dairies and milk-shops. Led to the 1879 Dairies, Cowsheds & Milkshops Order.

1879	Dairies, Cowsheds and Milkshops Order	Cowsheds to have adequate lighting, ventilation, cleansing, drainage and water supply to satisfaction of local authority. Milk from diseased animals could not be used as human food.
1885	Dairies, Cowsheds and Milk-shops Order	Provided further structure for local authorities to register dairymen, cowkeepers, and milk purveyors. Specified dairy regulation, providing guidelines for the lighting, ventilation, cleansing, drainage, and water supply of dairies and cowsheds.
1899	Sale of Food & Drugs Act	Led to the Sale of Milk Regulations (1901) that specified minimum level of fat in milk.

The Liverpool cowkeepers responded both individually and collectively. Firstly, as they were now required to have their premises inspected, they decided to throw open their doors to the general public so that their customers could see for themselves the high levels of cleanliness and hygiene that were practised. The phrase 'Inspection Welcome' became a common mantra and appeared on signs, hoardings and business stationery. Secondly, they decided to get themselves organised into a professional body and put on a show – an agricultural show.

Agricultural shows came about due to the need to develop better breeds of cattle. They have their roots in the latter part of the eighteenth century, during the reign of (Farmer) George III. Jenkins (1970) describes how one of the chief problems of livestock farming had been feeding during winter – most cattle were slaughtered and the meat salted in the autumn. It was only after the crop experiments of Lord 'Turnip' Townshend and Richard Weston that this extraordinary waste was avoided. Townshend grew turnips and Weston grew clover for feeding cattle throughout the winter. As a consequence of this innovation, valuable animals could be saved and, for the first time, farmers had the opportunity to improve the quality of their stock through breeding.

The formation of the Board of Agriculture in 1793 helped pool the knowledge and experience of farmers for the benefit of all. In 1839 the first show took place run by the English Agricultural Society and herd books were created to record the pedigree of the country's best breeding animals. The British Dairy Farmers' Association was founded in 1876 and, that year, held its first annual dairy show in the Agricultural Hall in Islington. Agricultural dairy shows became institutions.

Initially, the Liverpool cowkeepers had the opportunity to participate when either the national show or the county show came to the city. In Lancashire, which at that time included Liverpool, the county show was organised by The Royal Manchester, Liverpool and North Lancashire Agricultural Society (later, in 1893, to become The Royal Lancashire Agricultural Society). Shows were held in towns all over the county, including Manchester, Oldham, Blackburn, Burnley, Bury, Rochdale, Southport, Blackpool, and Lancaster. The county show came to Liverpool on a number of occasions: 1849, 1852, 1859, 1871, 1874, 1883, 1892, 1893, 1899, 1905, 1914, 1931, 1923 and 1938.

The county show did not take place in those years when the national show was held in Lancashire. The national show – run by the Royal Agricultural Society of England – came to Liverpool in the following years: 1841 (Falkner's Field – 7 acres), 1877 (Newsham Park – 75 acres) and 1910 (Wavertree Playground – 108 acres). These shows were very popular – the 1910 show experienced 138,000 attendances.

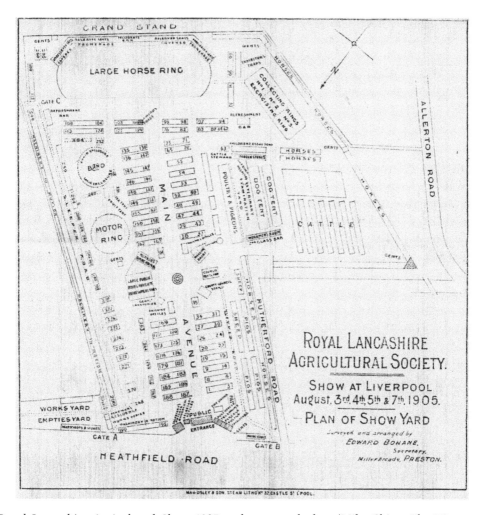

Royal Lancashire Agricultural Show 1905 – showground plan. (Mike Chitty, The Wavertree Society)

The Royal Agricultural Show, Wavertree Playground, 1910. (Mike Chitty, The Wavertree Society)

The Royal Agricultural Show, Wavertree Playground, 1910. (Mike Chitty, The Wavertree Society)

However, it seems that these occasional opportunities to show their stock were not frequent enough for the Liverpool cowkeepers. They decided to get themselves organised and then to hold their own show on an annual basis.

The *Liverpool Mercury* of 29 September 1865 reported on a 'Meeting of Cowkeepers And Dairymen' that had taken place the evening before. The meeting was held at the Oddfellows' Hall in St Annes Street, with the aim of establishing an association for insurance against the loss of cows by disease or accident and for the regulation of the price of milk. The association would be called 'The Liverpool Cowkeepers Association' and membership would be open to all persons keeping cows in the city. At the time, Liverpool had 338 cowkeepers, owning 2,460 cows.

A number of rules were discussed and agreed upon. Membership would be by annual subscription following the payment of a five shilling entrance fee. Any person desiring to become a member had to have his stock visited by a veterinary surgeon appointed by the association. It was suggested that two-thirds of the value of a cow be paid to the owner in the event of loss, instead of the full amount.

The meeting also discussed the price of milk. It was felt that the low price of 3*d* per quart was one reason why milk was being adulterated by adding water in order to increase profit. An increase to 4*d* per quart was agreed on the basis that the customer would support such an increase if it meant better-quality milk. The price increase would be implemented from 16 October.

A letter to the editor of the *Liverpool Mercury* was reproduced in the edition of Monday 31 October 1881. The letter was from 'An Old Cowkeeper', and it called upon the Haymarket Farmers Club and the committee of the Cowkeepers Association to organise a Christmas dairy show at the Haymarket. In making his case, the old cowkeeper emphasised the interdependence between the cowkeepers and the farmers, comparing the relationship to 'what a flower is to a bee'. This was a reference to the extensive trade that took place between farmers on the edge of the city and the cowkeepers within the city. The farmers would come to the Haymarket to sell hay, straw and feed to the cowkeepers and they would return to the farm with a cartload of best-quality cow muck, which the city cowhouses produced in great quantity.

Whether or not this anonymous letter had any influence on the matter, the following year the Liverpool and District Cowkeepers Association organised its first Christmas show of dairy cattle. The show took place at Mr Thomas Carr's recently opened auction yard in Breck Road, on the morning of 15 December 1882. The show was well attended, with 107 entries for the following list of classes:

Open classes:
Class 1 – Fat Cow, 40 score and upwards.
Class 2 – Fat Cow, over 32 and under 40 score.
Class 3 – Fat Cow, not exceeding 32 score. (G. Joy, Wavertree – 3rd place)

Classes limited to members of the association:
Class 4 – Fat Cow, 40 score and upward.
Class 5 – Fat Cow, over 32 and under 40 score.
Class 6 – Fat Cow not exceeding 32 score.

Class 7 – Cow in calf or milk, heavy weights.
Class 8 – Dairy Cow, light weights.

In addition, special prizes were awarded for Best Animal, Best Fat Animal and Best Dairy Cow.

After the show, members of the association met at the nearby Seaman's Hotel for dinner. During the speeches it was said that the show had demonstrated to the public that the cows kept in the dairies of Liverpool were in excellent condition – in fact, a better class of fat dairy cow had never been seen. In the afternoon some of the prize cattle were sold at Mr Carr's yard.

The Christmas Fat Cow Show became an annual event and proved to be extremely popular. So much so that in 1886 it was moved to the larger premises known as North Haymarket.

At the 1888 event, the after-dinner speeches touched on recent criticism of the quality of milk from the city's cowkeepers. Mr Webster, Chairman of the Farmers' Association, said that as a rule country milk was thought to be much better than town dairy milk; however he and his associates knew it not to be so. The farmers knew where their best hay went – to the dairymen of the town – and they also knew that the best hay would produce the best milk.

Mr George Verity, in proposing the toast of the Liverpool and District Cowkeepers Association, said that seven or eight years ago the association had no intention of having a fat and dairy cattle show, and it was not until many gentlemen of high standing – who knew nothing of cattle – began to criticise and say that there was an unhealthy lot of cattle in the Liverpool shippons that they determined to hold the show. They had the result of their show that day to disprove these adverse criticisms. Without cows being kept in the city, the people of Liverpool would be dependent upon milk that was not less than twelve or twenty-four hours old. Now they had milk delivered morning and evening and this was very important considering that scores of children were brought up entirely on cows' milk.

Mr F. Stoner replied, saying that the cowkeepers of Liverpool held this show to bring before the people of Liverpool the real nature and quality of the cattle that had been so unjustly criticised. The milk produced in Liverpool was far better and purer, and there was less contamination in it than a good deal of the milk produced in the country. During the whole of 1887 there had not been a single case of disease among the cows in the shippons of Liverpool.

Having their own annual event did not prevent the Liverpool cowkeepers from participating when the county or national shows came to town. Indeed, the Liverpool and District Cowkeepers Association would put up prize money for a number of classes that were only open to entrants from within the association area:

- Town Shippons and Dairies – competition open to all Cowkeepers within a radius of seven miles of the Town Hall.
  - Class 26 – Best Kept Shippon, Milkhouse and Shop or Dairy. *General Appearance, Arrangements for Convenience, Economy of Labour, Cleanliness, Air Space, Light and ventilation of Shippon, has been taken into account in awarding the prizes.*

- *Town Dairy Cows* – competition limited to Members of the Liverpool & District Cowkeepers Association (and kept on premises within the district of the association at least three months after the show).
  - ○ Class 24 – Town Dairy Cow, any breed or crossbreed, in milk or in calf, over 11 cwt.
  - ○ Class 25 – Town Dairy Cow, any breed or crossbreed, in milk or in calf, not exceeding 11cwt.
  - ○ Class 93 – Cowkeepers Milk Turn-out over 14 hands.
  - ○ Class 94 Cowkeepers Milk Turn-out not exceeding 14 hands.

Success at these shows gave the prize-winners a unique window of opportunity (until the next show) to publicise the quality of their premises, practices and products. Signs and banners were erected proclaiming the title of 'Best Kept Shippon' or 'Best Cow' of whatever class. Certificates and cups were displayed in prominent places on dairy walls or shop windows. Prize-winning cattle were paraded around local streets wearing their rosettes. All this pageantry took place with one principal aim in mind – that of gaining a competitive advantage over other dairies, but especially over those dairies selling railway milk.

For the 1883 county show, the association also donated a prize for the best essay on the mode of Liverpool cowkeeping. The prize was won by Frederick Stoner of Great George Street, Liverpool. This essay gives a snapshot of the scale of cowkeeping in the city at that time, afforded by the recent introduction of compulsory registration as required under the Dairies, Cowsheds and Milkshops Order of 1879. By 1882, of the 1,577 premises registered for the sale of milk, 432 were also licensed to keep cows.

The table below illustrates the number of shippons and cows that were licensed during the period 1879–1882.

The standards were demanding and their enforcement was rigorous. Consequently, not all applications for licenses were successful. When the order first came into force in 1879, of 2,099 applications received, 876 were refused.

Of course, this drive for quality was not the sole domain of the authorities; it was as much a cause for the cowkeeping profession. By joining an association, cowkeepers not

The Dairies, Cowsheds and Milkshops Order 1879 Licenses Issued In Liverpool (1879–1882)		
Year	No. Licensed Shippons	No. Licensed Cows
1879	452	5,235
1880	450	5,322
1881	443	5,460
1882	432	5,684

only gained professional status, but also the credibility of meeting the association's own exacting standards of husbandry.

Moreover, for the Dales folk who had relocated to the city, the Liverpool and District Cowkeepers Association also had an important social function. Most ethnic/cultural minority groups that migrated to Liverpool tended to dwell in a fixed geographical location within the city (e.g. 'Chinatown'). In this way they supported each other and maintained their community and cultural identity. However, clustering in a fixed area was not an option for the cowkeepers; they needed to spread out in order to be near their customers. So, the regular meetings of groups such as the cowkeepers association or the young farmers club became a way of maintaining contact and sharing a common culture. They were more than a place for discussing the relative merits of a polley, a blue-grey, or a white-face – the Liverpool and District Cowkeepers Association had an extensive social programme, including dances and day trips. And it was at such events that young cowkeepers would meet their future spouses.

From Wharfedale to Liverpool

By the time of the census of 1861, George Joy, his brother Orlando John, and his sister Hannah were living at 77 Lovat Street in the Wavertree district of Liverpool. Both George and Orlando were given the occupational description of 'cowkeeper'.

Wavertree is located to the south of the city centre and at that time would have been on the outward edge of the expanding city population. This location may have provided better access to grazing pasture than was available in the inner city.

The two brothers and one sister business combination seems to have been a success and by 1863 they were developing a satellite business in Garston – at that time, a rural township lying to the south of the city. However, on 22 February 1866 Hannah married. Connections with the family's Yorkshire roots had obviously remained intact, for she married Thomas Stockdale of Rainlands Farm in Hebden. Although the ceremony took place in St John the Evangelist Church in the Knotty Ash district of Liverpool, Hannah returned to the Dales to live with her husband and his family at Rainlands Farm.

Some six months after his younger sister had entered wedlock, George followed suit. However, he married locally and remained in Liverpool. On 2 August 1866 George's marriage to Mary Ann Maria Spracklin was solemnised in the parish of Everton, Liverpool. When his first daughter was born, in November 1867, George was no longer living at 77 Lovat Street. He had set up home and business at 88 Ash Grove, still in the district of Wavertree but a little further south.

Orlando remained at Lovatt Street. However, it seems that in-laws from Yorkshire became involved in running the business. By the time of the 1871 census, Thomas and Elizabeth Hardacre, also from Hebden, were in residence with Orlando (Orlando's older sister, Mary Ann, had married Jeramiah Hardacre. Thomas Hardacre was Jeramiah's brother). The three may well have had more than one property between them.

Orlando's continued Yorkshire connections were underlined when on 6 April 1876 he married Hannah Daggett. They also were married at St John the Evangelist Church, in the Knotty Ash district of Liverpool. At that time, Hannah's parents were back home in the Dales, farming at Rams Close. But her brother, John William Daggett, was living with his family at 20 Duke Street in Garston, earning a living as a cowkeeper. Orlando and Hannah were living at 20 Duke Street when they had their first child, Elizabeth (Betsy) Hannah, on

Orlando John Joy with daughter, Betsy Joy, *c.* 1880. (Author)

27 June 1876. However, by the time that their second child, Elsie, was born, in March 1878, Orlando had moved back to Yorkshire and was living in Ilkley. Both daughters were christened at St Wilfrid's Church, Burnsall on 6 October 1878 and the parish register shows that Orlando had reverted to a previous family occupation – that of innkeeper. He spent the rest of his days as the proprietor of the Midland Hotel, Station Road, Ilkley.

In 1870, George was listed as a cowkeeper working out of 2/4 Godfrey Street in Everton. However, his main residence was still at Ash Grove. In 1873, he placed the following – by today's standards, politically incorrect – advert in the *Liverpool Mercury* of Monday 17 November:

> WANTED – a servant girl of all work, one who can milk. A Protestant. Apply to Mr Joy, Ash Grove, Wavertree.

By the time of the 1881 census, the family living at Ash Grove consisted of George, his wife, his father-in-law and six children: Mary E. Joy (age thirteen), Flora S. Joy (age eleven), George A. Joy (age nine), William Joy (age eight), Edward Joy (age four), Tom S. Joy (age one). The Ash Grove property was not quite the typical cowhouse in that it was a detached property, albeit on the end of a terraced row. The building stands today and shows evidence of previous extension, together with having a side/back yard large enough in which to keep cows.

As was the case with many Dales families, George had cousins who had followed him out to Liverpool. Margaret Joy (daughter of Anthony and Maria) had married John Metcalfe in 1868. Three years later the couple were running a cowhouse at 20 Luke Street, but by the time of the 1881 census they had moved to 2A Ash Grove and were keeping cows at the other end of the street from George and his family; they would later move to 8 Ash Grove, raise a family and spend the rest of their days as cowkeepers. Also, Margaret's sister,

88 Ash Grove, Wavertree, 2014. (Author)

Hannah Maria, had married George Brown and for a while they kept cows at 2 Lully Street (1871 census) before returning to Yorkshire to continue farming there.

George and his family remained at Ash Grove up until at least the 1891 census. By 1901 the family had relocated to Calton Dairy, on the corner of Smithdown Road and Calton Avenue (one road along from the now famous Penny Lane). This property had access to a shippon on Smithdown Road that was spacious enough to house a large herd of cattle. George would have been in his seventies and his wife nearing her sixties. Under these circumstances the support of their family would have been crucial to the continuation of the dairy business. Their daughter Mary E. Joy (age thirty-three) and their youngest sons, Edward Joy (age twenty-four) and Thomas S. Joy (age twenty-one) were still living in the family home and were no doubt taking a bigger role in the running of the business.

Former Calton Dairy, Calton Avenue, Wavertree, 2014. (Author)

At the time of his death, on 30 May 1904, George's address is given as 21 Calton Avenue – the property adjacent to the dairy. The executer of George's estate was his son, Edward Joy, whose occupation is still given as 'cowkeeper'. George was buried on 2 June 1904 at Holy Trinity Church in Wavertree.

George's two remaining sons, Edward and Tom, continued the family business. At the county show held in Liverpool in 1905, there were numerous prizes won by the Joy brothers:

Class 18 – Dairy Cow in Milk (any weight). Reserve [4th] – Joy brothers, 362 Smithdown Road, 'Nancy'.
Class 20 – Dairy Cow in Milk, over 11cwt. Reserve [4th] – Joy brothers, 362 Smithdown Road, 'Beauty'.
Class 23 [Town Dairy Cows] – Dairy Cow in Milk or calf over 11cwt. Highly Commended – Joy bothers, 362 Smithdown Road, 'Nancy'.
Class 25 – Pair of Dairy Cows in Milk or Calf, any weight. Reserve [4th] – Joy brothers, 362 Smithdown Road.
Class 113 – Cowkeepers Milk Turn-out, 14 hands or more. Commended – Joy brothers, Carlton Farm Dairy.
Class 114 – Cowkeepers Milk Turn-out, less than 14 hands. Reserve [4th] – Joy brothers, Carlton Farm Dairy.

The 1911 Gore's Directory lists the Joy brothers as still living at 21 Calton Avenue, but also as working out of 362 Smithdown Road, located at the top of Calton Avenue. It was common practice for corner properties to be listed twice in directories. Today, the original house on Calton Avenue has been demolished and replaced. However the business premises on Smithdown Road are still in existence and are now occupied by a Halfords Autocentre.

Former shippon – 362 Smithdown Road, 2014. (Author)

It seems that both brothers continued to keep cows in and around the Penny Lane area. When Tom passed away in 1918, his occupation was still being given as cowkeeper and he was living at 1a Olivedale Road. In the 1938 Kelly's Directory, Edward is listed as a dairyman at 12 Carsdale Road. Both of these properties are end-terraced buildings and may have been used for the sale of milk, but it is not clear whether or not they were used for the keeping of cows. Edward died in 1950.

It is interesting to note that each time George relocated the family business within the city of Liverpool, he moved further away from the city centre. This outward migration of George's successive dairies and cowhouses took place over a period of some sixty years and may have reflected the growth of the city during that time. As the city expanded, it gradually absorbed any open land that might have previously been available as pasture. This wave of urbanisation would have pushed before it any cowkeeper who wished to continue grazing his cows. The table below shows the approximate distance from the city centre (town hall) of each of George's successive properties.

The migration of George Joy's cowhouses in Liverpool 1861–1911

Census Year	City of Liverpool Population	Street Address of George Joy's Dairy Premises	Distance from City Centre (miles)
1861	429,881	77 Lovat Street	1.86
1871	539,248		
1881	648,616	88 Ash Grove	2.61
1891	659,967		
1901	711,030	362 Smithdown Road	3.51
1911	766,044	21 Calton Avenue	

12 Carsdale Road, Wavertree –
abode of dairyman Edward Joy.
(Author)

1a Olivedale Road, Wavertree –
abode of Tom Joy, Cowkeeper.
(Author)

From Wharfedale to Garston

Initially, Daniel Joy, the brother of George, Orlando and Hannah, remained in Yorkshire and made a living by combining farming with innkeeping. Indeed, he married the daughter of an innkeeper and together they made a life for themselves and their eight children:

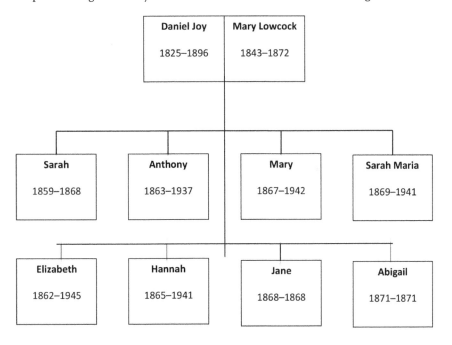

To begin with, they lived at Bolton Bridge and Daniel worked as a stockman at the nearby Devonshire Arms. He obviously took his role as stockman very seriously, as reported in 1866 by both the *Leeds Mercury* (19 November) and the *Bradford Observer* (22 November) when covering the impact of the great floods that took place that year:

At Bolton Bridge the flood expanded over the meadows, and the two houses not far from the Red Lion were inundated, though not to any serious extent. Daniel Joy, who resides near the

Devonshire Arms, barely 'saved his bacon' by securing his pigs when in extreme danger of being washed away.

However, tragedy was to befall the family and, after losing three daughters to child- and birth-related illnesses, Daniel's troubled wife, Mary, took her own life in the December of 1872. As well as coping with his own grief at the sudden and tragic loss of his wife, Daniel had to find a way of caring for his very young family of five, consisting of: Elizabeth (age ten), Anthony (age eight), Hannah (age seven), Mary (age five) and Sarah Maria (age two). Clearly, it was not going to be easy for him to be both the main breadwinner and carer. In such circumstances the natural thing to do would be to look to your extended family for support.

Daniel did have options and could have remained living in Yorkshire. Perhaps, though, it was the need to draw a line and move on from the tragedy that was the death of his wife and mother of his children that made Daniel decide to make a complete break by following his brothers in exchanging the banks of the Wharfe for the banks of the Mersey. Daniel and his young family relocated to Garston in 1873, less than a year after Mary's death. There, Daniel took over the running of a dairy that had been established by his brothers, George and Orlando.

At that time Garston was not part of the city of Liverpool, but rather was a township and a port in its own right, separate from and lying to the south of the city. The original village of Garston had evolved near the point where the Garston River ran into the Mersey. It had seen some development as a port town, but the big change came in the early 1850s when the St Helens Canal & Railway Company built an enclosed dock and serviced it with substantial railway sidings (see 1850 map across). This new dock became a magnet for important national industries in need of an export-import gateway for raw materials and manufactured goods.

The port was so successful that in 1867 a second 'New Dock' (eventually to be called 'North Dock') was built (see 1894 map on p. 54). The combination of docks, railway links and available land was irresistible and the port became home to a variety of industries

Junction of Church Road and Railway Street, Garston, c. 1900. (Garston & District Historical Society)

A 1988 view from church tower showing Railway Street (right) and, in the background, dock land that was formerly fields of 'Dale Farm'. (Garston & District Historical Society)

including bobbin and shuttle manufacturing, bottle works, copper rolling, various iron and metallurgical works, distillery, tannery, ship yard, saw mills and banana importers. With these new industries, the town of Garston was expanding. In 1841 it had a population of 1,888 but by 1871 that figure had swelled to 7,840.

Daniel moved his family to dairy premises in Railway Street, off Church Road. It was located close to the docks and the railway (where the railway crossed over Church Road) and was opposite Garston Church. He was able to keep his dairy herd at Dale Farm, which ran down to the shore, and sell milk from his home in Railway Street.

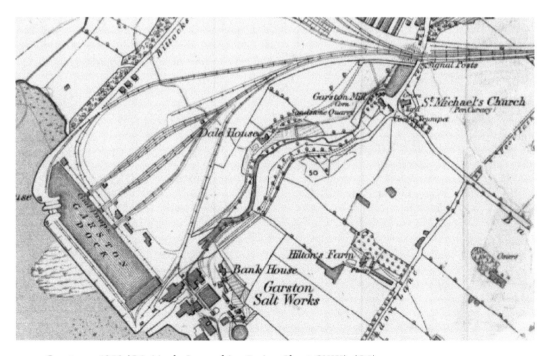

Garston – 1850 (OS 6-inch. Lancashire Series, Sheet CXIII). (OS)

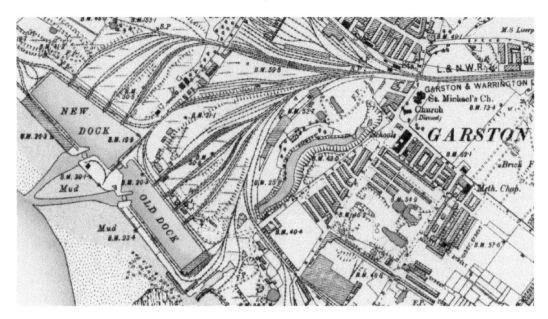

Garston – 1894 (OS 6-inch. Lancashire Series, Sheet CXIII.SE). (OS)

The church at Garston was already a focal point for the local community and it was to become an important part of the lives of Daniel and his family. When Daniel arrived in Garston, plans were already afoot to build a new church, what would then become the third church on that site. This was a massive undertaking and the top of Railway Street would have been a hive of construction activity for the next five years. When the new church was consecrated in 1877, the whole of the Joy family were in attendance, with Daniel's son, Anthony, singing in the church choir.

It seems that by 1878 Daniel's business was well and truly up and running. He was not only selling milk but was also doing a tidy sideline in selling manure to local farmers. Daniel's memorandum book contains a note written in his own hand. It is a record of loads of manure sold to a Mr Atherton (possibly of Chapel House Farm, Speke). The account commences on 10 May 1878 and runs up until 2 January 1879. During that period Daniel sold twenty individual loads of manure to Mr Atherton.

There are two relevant records in the 1881 census. The first record gives Daniel's place of residence as 7 Railway Street in Garston, Liverpool. His age is given as fifty-three and his household includes three daughters: Hannah (age sixteen), Mary (age fourteen) and Sarah M. (age eleven). The second record is of the family of George Joy at 88 Ash Grove, Wavertree. Recorded along with George's immediate family are also his niece Elizabeth Joy (age nineteen) and his nephew Anthony Joy (age seventeen). It is not clear whether or not Daniel's two eldest children were living with their uncle or were just visiting there at the time of the census; either way, they would have been actively helping George's family in the running of the business.

Daniel's children all married locally. On 25 October 1884, Daniel's eldest daughter, Elizabeth, married Thomas Caleb Stringfellow of Christchurch, Crewe. Four years later, in 1888, there followed a spate of matrimonial activity in the family. On 17 May, Anthony

Account of manure sold by Daniel Joy to Mr Atherton, 1878/79. (Author)

married Ann Jane Percival at St Michael's Church, Garston. One month later, on 20 June, Hannah married George Wright, a railway fireman, of 15 James Street, Garston. And then, on 24 December, Mary married William Yoxall of Wilmslow, Cheshire at St Cleopa's Church, Toxteth.

Once all but the youngest of his children were happily married, Daniel decided to make provision for his own demise and he bought a burial plot in the newly enclosed graveyard at St Michael's Church, Garston. The plot he chose was at the south-west corner of the site, overlooking the fields of Dale Farm where he had spent so much of his working life since moving to Garston. It is not known exactly when Daniel had to quit Dale Farm, but it is likely that the ever-expanding port needed whatever space was available. Ultimately, the land would be used for the construction of Garston's third dock, Stalbridge Dock, together with its attending railway infrastructure.

By 1891, Daniel (age sixty-five) was living at 10 Railway Street. Living with him were his daughters Sarah Joy (twenty-one) and Mary Yoxall (twenty-four) together with his son-in-law, William Yoxall. It is probable that his two daughters and one son-in-law were actively involved in the running of Daniel's milk business. The Kelly's Directory of 1895

Joy family grave, St Michaels Church, Garston – overlooking the former site of 'Dale Farm' in 2014. (Author)

includes, among the list of dairymen, an entry for Joy, Daniel, 10 Railway Street. Indeed, Daniel was a cowkeeper right up until the very end. He passed away (senile bronchitis ischemia) on 3 February 1896, aged seventy. His effects of £140 were administered to his son, Anthony Joy, also a cowkeeper.

It was not only Anthony who kept up the family tradition of cowkeeping; two of Daniel's daughters and their respective husbands followed suit. Firstly, Mary and her husband, William Yoxall, established a cowkeeping business at 26 Willoughby Street in the Edge Hill district of Liverpool (Kelly's 1894 and Gore's 1900 directories of Liverpool). Also, by the time of the 1901 census, Hannah and her husband, George Wright, were running a cowkeeping business out of Dale House at the top of Dale Street in Garston. Ten years later, George was a milk dealer and Hannah a confectioner, living at 262 Garston Old Road. Their son, George Eric Wright, was employed in the family business as a 'milk dealer's assistant'.

In 1912 George Eric Wright emigrated to Australia and worked as a dairy farmer on the Maryland Estate, located in the Bringelly district of Sydney, New South Wales. With the outbreak of war, he enlisted with the Australian Imperial Forces and fought at Gallipoli and in Egypt. On 24 July 1916, he was killed in action during the battle of Pozieres in the Somme Valley. He was twenty-six years old.

Twentieth-Century Milk Wars: Politics and Pasteurisation

At the beginning of the twentieth century, Britain was still wrestling with what the Victorians had referred to as 'the milk problem'. The problem was complex. One difficulty was trying to standardise a product that by its very nature varied from place to place, from farm to farm and from cow to cow – and even from the *same* cow – according to conditions at the time. The science for identifying adulteration or contamination had not been exact and was subject to extensive debate – in particular, with regard to whether milk was or was not the cause of disease.

Identifying the source of adulteration or contamination was difficult in a long supply chain that might include cow, milker, dairyman, collector, transporter, distributer, retailer and customer. At a ministerial level, responsibility seemed to fall between the two (milking) stools of Health and Agriculture, with experts in both of these camps vying for authority. And, in terms of the market, the small producer-retailers were still battling for survival against the ever-expanding corporate dairies, with both parties actively lobbying their respective political representatives accordingly.

As the science gradually improved, so it became possible to better describe milk in terms of both its chemical composition and its bacteriological content. The former led to a call for the introduction of a grading system for milk and the latter for compulsory pasteurisation. One was to succeed, the other to fail. Both were to play a part in the legislative framework affecting milk that developed in the first half of the twentieth century.

Tuberculosis

Edward William Hope was the Medical Officer of Health for the City and Port of Liverpool from 1894 until 1924. In his book *Health at the Gateway – Problems and International Obligations of a Seaport City* (1931), he puts into historical context the health issues that were peculiar to the city of Liverpool.

As a busy port, Liverpool had developed a reputation as being 'The Gateway to Europe'. However, as well as bringing wealth, the city's success as a port also brought threats to its

Year	Legislation	Impact
		Twentieth-Century Legislation Affecting Liverpool Cowkeepers
1900	Liverpool Corporation Act	Empowered the Corporation to inspect farms in other areas that were supplying milk to the city.
1901	Sale of Milk Regulations	Stipulated that milk should contain at least 3 per cent milk fat and 8.5 per cent other. There were no minimum standards of cleanliness.
1902	Liverpool Corporation Act	Stanley cattle market acquired by Liverpool Corporation from the Liverpool New Cattle Company.
1917	Milk (Special Designation) Order	Introduced grading.
1918	Milk (Prices) Order	Recognised grading.
1918	Trade Boards Act	Defined hours and conditions of labour and, through the 1920 Trade Boards (Milk Distributive) Order, applied these to the sale of fresh milk and all its associated operations.
1920	Trade Boards (Milk Distributive) Order	
1921	Liverpool Corporation Act	Premises for the keeping of cattle had to be licensed for this purpose by Liverpool Corporation and would be subject to inspection by officials of the Corporation. Also, the keeping of cows on unlicensed premises was subject to penalty.

1922	Milk and Dairies (Amendment) Act	Outlawed the sale of tuberculous milk and introduced licensed designations, including 'Pasteurised' through the 1923 Milk and Dairies Order, which gave detailed definitions of graded milk.
1923	Milk and Dairies Order	
1925	Milk and Dairies Act	Methylene Blue test was officially adopted for raw, graded milks. Updated the regulations about buildings and equipment. The sterilisation of milk vessels and appliances was required. Provisions about animal health, allowing a potentially diseased source of milk to be stopped. All producers and retailers of milk now had to be registered. The cooling of milk was made compulsory.
1926	Milk and Dairies Order	
1936	Milk (Special Designations) Order	Prescribed five new grades of milk replacing 'Grade A'.
1937	Agriculture Act	Enabled the Ministry of Agriculture's veterinary inspectors to examine and test herds holding Accredited licences.
1944	Food and Drugs (Milk and Dairies) Act	Shifted responsibility for cleanliness to the Ministry of Agriculture.
1949	Milk and Dairies Act Milk and Dairies Regulations	Insisted for the first time that all milk had to be graded.
1949	Milk (Special Designations) Act.	Introduced first local option of compulsory pasteurisation of milk.

health. Many of the immigrants arriving at the docks were destitute and fever-stricken. In response, a combined voluntary and municipal effort was stimulated to tackle the city's health problem. Consequently, the city of Liverpool became a pioneer in methods for the control and treatment of disease. Indeed, on many occasions the Corporation of Liverpool promoted sanitary legislation, which, after successful rehearsal in the city, was then incorporated into Public Health Acts.

The first appointments of a medical officer for health, of sanitary officers, of a city bacteriologist and of women health officers, were made in Liverpool. It was also in Liverpool that voluntary effort established the framework for district nursing, the first welfare clinics and the Society for the Prevention of Cruelty to Children. Liverpool also became the home of the country's first School of Hygiene.

The Public Health Act of 1872 authorised the establishment of port sanitary authorities. In Liverpool, where the prevalence of infectious disease was greater than elsewhere in the country, it had long been held that better organisation of port administration, medical services and hospital accommodation was necessary. With the intention of delivering this, the Council of Liverpool became the port sanitary authority.

One of the many diseases affecting the city was tuberculosis. In the 1870s, tuberculosis of the lungs was killing 4.3 per 1,000 of the population per year. Victims were mainly breadwinners, leaving young families destitute. A breakthrough came in 1882 when Robert Koch was successful in isolating the *tubercle bacillus* responsible for the disease. After that there was extensive discussion on the many aspects of the disease.

In 1895 a royal commission expressed its considered opinion that the consumption of tuberculous milk was a contributory factor to human tuberculosis. A year later the city council engaged expert bacteriologists to examine and report upon samples of milk being sold for human consumption in Liverpool. The results showed that tuberculous milk was being sold; however while 5 per cent of that being sold by city cowkeepers was found to be infected, of the milk being brought into the city from the surrounding countryside, a staggering 13.4 per cent was infected. This led to the city council applying to Parliament for special powers to visit and inspect country cowsheds supplying milk to Liverpool. Using its powers under the 1900 Liverpool Corporation Act, city inspectors discovered that sanitary conditions in cowsheds in the country were generally much inferior to those found in the city cowsheds.

In order to obtain more information about the extent and location of the disease within the city, in 1901 Liverpool inaugurated a voluntary notification scheme. This brought the sufferer into direct contact with the health authority, which provided a card containing simple instructions about their mode of living. On average 2,300 cases were reported each year until 1908, when the scheme was rolled out across the country by the local government board.

As well as sanatoriums being established across the city to house and treat those infected, exhibitions and lectures were arranged to help increase awareness of the disease; these proved very popular with the general public. No less than 40,000 people attended one such exhibition, held for a week in the summer of 1910 in St Martin's Hall, Scotland Road. Among the special-interest groups attending these events was the Liverpool and District Cowkeepers Association.

Of particular relevance to the city cowkeeper was the recognition of the tuberculous-carrying cow. Early efforts to tackle the neglected condition of cowsheds had been based on general methods of sanitation without suspicion that the milk of the tuberculous cow was the source of the disease in the human consumer. Now, with greater understanding of the nature of the disease, the cowkeepers began to work more closely with the health authority and co-operated in the development and implementation of new regulations and improved practices.

Some of this work was experimental in nature. Between 1911 and 1913, under the stewardship of Alderman Anthony Shelmerdine JP, Liverpool City Council investigated methods of treating milk with electricity! This research was carried out at the council's Earle Road milk depot. One method, that of using a rapidly alternating current at high potential, had a degree of success and was introduced and operated in Liverpool for a number of months.

By working closely with the Health Department, the Liverpool and District Cowkeepers Association played its part in reducing the number of tuberculosis-related deaths in the city, from 15,572 (350.7 per 100k *p.a.*) in 1865 to 1,058 (121.2 per 100k *p.a.*) in 1929.

Pasteurisation

Elsewhere, there was one other new practice that was proving to be more successful, but also to be quite contentious and even divisive within the national milk industry. That practice involved the heating or 'pasteurisation' of milk and the question of its introduction developed into an issue that would embroil all city cowkeepers.

In his paper 'The Pasteurisation of England' (2000), Atkins describes how the adoption of pasteurisation as a technological answer to tuberculosis was slow at first. Indeed, although the first commercial pasteuriser was manufactured in Germany in 1880, in Britain the majority of retail milk was still unpasteurised by 1939.

One reason for this was that the early pasteurisation machinery was unreliable. The problematic flash method (in which milk was heated very quickly to a high temperature in batches and then cooled) was eventually banned by the Milk (Special Designations) Order (1923) and replaced for the next couple of decades by low-temperature (63–71°C for thirty minutes) machines, which heated the milk slowly as it passed through a succession of large vessels. Eventually, High Temperature-Short Time (HTST) methods were introduced in the 1940s, whereby the milk was heated for 12–20 seconds at 75–76°C.

As would be expected, the large dairy companies, always with one eye on the possibility of eliminating their smaller competitors, were in favour of pasteurisation. The process fitted in well with their scientific approach and, although there would be a cost, the size of their operation would deliver economies of scale. Furthermore, the process would extend the life of their milk by offsetting the souring process while the milk was in transit. They also supported wartime efforts to rationalise delivery rounds, which would have been to the detriment of the producer-retailers.

The cowkeeping producer-retailers fought back in their opposition to pasteurisation. Many could not afford their own pasteurising and bottling plant and would have had to take their milk to a depot and receive some amalgamated product in return – the direct and special link with the cow, upon which much goodwill depended, would have been lost. Their opposition to the move towards the compulsory pasteurisation that was implicit in the Milk Industry Bill of 1938 contributed to it not reaching the statute book. Immediately after the First World War, just when pasteurisation was becoming more common in the larger cities, science presented the anti-lobby with its best argument yet – that milk contained micronutrients. The case against pasteurisation was then that the process might destroy these beneficial vitamins, trace elements, enzymes, antibodies, and hormones.

However, although the anti-lobby was successful in preventing pasteurisation from becoming compulsory, they could not prevent the pasteurised condition being recognised as a class of milk in the various grading frameworks that evolved and were introduced during the 1900s.

The grading of milk

The grading and the certification of milk were two of the principal requirements of the clean milk movement, which took a significant leap forward with the founding of the National Clean Milk Society in 1915. The society's stated aim was to raise the hygienic standard of milk and milk products and to educate the public as to the importance of a clean and wholesome milk supply.

There was competitive lobbying by different segments of the milk trade and this may have contributed to the fragmented response from government. While the Astor Committee was still considering the matter, the ministries of Food and Health acted unilaterally and introduced a simple grading system (Grade A and Grade B) in the Milk (Special Designation) Order (1917). The Milk (Prices) Order (1918) also recognised this grading and from September 1918 the Food Controller began granting licenses.

As these licenses were based on the local government board's scorecard for the inspection of dairy farms, they focused on the producers' premises. For example, to achieve Grade A, the milk had to be produced under exceptionally clean and hygienic conditions from a herd certified by a vet as free from tuberculosis and immediately bottled on the premises where it was produced. These premises had to score a minimum of 80 per cent for cleanliness of buildings, equipment and methods of production on the local government board's scorecard.

The cowkeepers continued to be suspicious of the motives of what they saw as the 'wealthy and elite' clean milk campaigners. They feared that they might be forced to invest heavily in special equipment and new buildings, which, being predominantly small operators in terms of their herd size, they could not have afforded. Their defiance gradually lessened as it became obvious that the standard for graded milk was within the reach of those who practiced the ordinary rules of cleanliness and that achieving the standard would also increase profits.

However, over the next fifty years the cowkeeper would find it more and more difficult to sustain their position in a rapidly modernising industry. As the science of milk became more effective and its processes more efficient, this was reflected in successive legislation that continuously changed and developed the system of grading. A summary of this evolution of milk grading can be seen opposite.

In 1944, responsibility for cleanliness shifted to the Ministry of Agriculture, as the evidence suggested that there was significant variation in the performance of different

Grades of Milk Specified in the Milk (Special Designation) Orders and Drinking-Milk Regulations	
Date	**Grades**
1917	• Grade A • Grade B
1923	• Certified • Grade A (Tuberculin Tested) • Grade A • Grade A (Pasteurised) • Pasteurised
1936	• Tuberculin Tested (Certified) • Tuberculin Tested • Tuberculin Tested (Pasteurised) • Accredited • Pasteurised
1949	• Tuberculin Tested • Accredited • Pasteurised • Sterilised
1954	• Tuberculin Tested • Pasteurised • Sterilised
1963	• Untreated • Pasteurised • Sterilised

local authorities. The ministry appointed a chief milk officer, with regional milk officers to inspect farms and give advice on buildings, equipment and production methods.

The 1949 Milk (Special Designations) Act and subsequent orders required for the first time that all milk had to be graded. By 1954 there were only three categories: Tuberculin Tested (TT) raw milk, which came from herds attested free from tuberculosis; pasteurised milk, which had to be sold in bottles; and sterilised milk. Once the national herd was declared free from tuberculosis, there was no further need for the TT designation and it was dropped in 1963.

Standing on the periphery of this revolution within the milk industry was the traditional city cowkeeper. For some, this landscape of change was too much to cope with and eventually they left the cowkeeping business, either to return to farming in the Dales or to take up alternative employment (utilising their stock-keeping skills wherever possible) in the city. Others embraced the changes, continuing the cowkeeping tradition but adapting it to meet the requirement of the new 'scientific' approach. The 1911 census indicates that the number of cowkeeping families living in Liverpool had fallen to 214. Moreover, those who remained still had their work cut out for them if they were to survive.

Wellington Dairy

With it being located at the end of his street, it is not surprising that Anthony continued the close association that his father had engendered with St Michael's, Garston Parish Church. Whether it was as choirboy, sidesman, lay rep or chairman of the church club, it was to be an association that would last his lifetime. Indeed, it was through the church that Anthony met his wife-to-be, Ann Jane Percival.

Anthony married Ann Jane at St Michael's Church on 17 May 1888. At the time of the marriage Anthony was living at 10 Railway Street with his father and sisters. At the ceremony the witnesses were Daniel Joy (cowkeeper) and Joseph Percival (farm bailiff). The

Anthony and Ann Jane Joy, 1925. (Author)

newly-weds moved into a house at 9 Jackson Street, just around the corner and a bit up the village from Railway Street. That meant that Anthony was still local enough to continue working at his father's dairy – as well as making a bit of extra money delivering coal.

Together, Anthony and Ann Jane had six children:

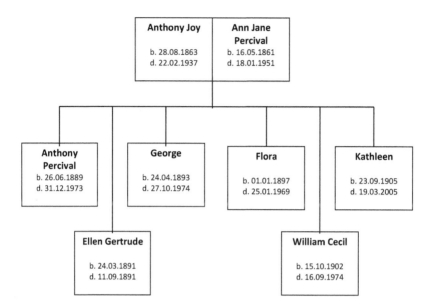

The Kelly's Directory of 1895 includes, among the list of dairymen, 'Joy, Anthony, 23 Island Road'. Clearly, by then Anthony was in the cowkeeping business in his own right; he was renting the property at an annual cost of £22 10s. By the time of Daniel's death in February 1896, Anthony's given address was the slightly modified 23a Island Road. This property was also known as Island Road Dairy.

The Island Road site was a typical cowhouse providing a tidy combination of shippon, yard and dairy and also had easy access to three large fields that were suitable for grazing. This land, just a hundred yards or so from the dairy (over the railway bridge), was once part of Island Farm owned by the Liverpool linen merchant Adam Lightbody.

However, the family's tenure at Island Road was to be relatively short lived. The continuing growth of the township of Garston was putting available land at a premium. By the end of the nineteenth century a more enlightened approach to town planning recognised the value of public open spaces for leisure and educational purposes. Garston Urban District Council was acutely aware of the lack of a technical school, public baths, outdoor recreation facilities and a modern library in the township. Consequently, the 35 acres bounded by Island Road, Long Lane, Whitehedge Road and Garston Old Road were acquired for public use. Garston Park (aka Long Lane Recreation Ground) was officially opened in 1901.

The Joy family were served notice to quit the land and were once again looking for suitable accommodation for their cowkeeping business in Garston. What they found was a more spacious property than Island Road Dairy. What they found was to become Wellington Dairy.

Former Island Road Dairy,
Garston, 2014. (Author)

Island Road leading to Garston
Park, 2014. (Author)

Anthony Joy, Wellington
Dairy – advert in parish
magazine, *c.* 1900. (Author)

ANTHONY JOY,

WELLINGTON DAIRY,

Wellington St.

The oldest established Dairy in
Garston.

The property at 37 Wellington Street was in two parts, seen on the plan below. The house, referred to as the 'Dairy House', was a two-up, two-down end terrace. To its rear it had a covered yard with an entrance off Duke Street. Leading off the yard were two outbuildings. One was a cold room, used for storing milk and also for making and keeping cheese. That single room was referred to as 'The Dairy'. The other outbuilding incorporated a washhouse and an outside lavvy. Running between the two outbuildings was a whitewashed passageway that led to the bottom of the yard. A gate at the bottom opened out on to the back entry that ran behind 37–43 Wellington Street, while access to the second part of the property was across the entry.

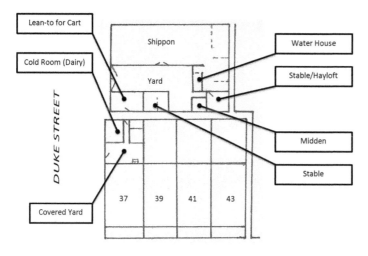

Plan of Wellington Diary.

Wellington Dairy – 37 Wellington Street, 2014. (Author)

The second part of the property consisted of an enclosed yard surrounded by a shippon, a water house, a hayloft, a midden, and stables. The yard had a set of large wooden gates that opened out onto Duke Street. When Anthony bought the property it was ready-made for a cowkeeping business.

In addition to these two adjoining properties, Anthony was also able to rent a number of fields off The Avenue. This was to enable his dairy herd to graze in open pasture. The cows would be brought down Woolton Road twice a day for milking in the shippon at Duke Street. Tom Ockleshaw recalls an occasion in the 1940s when, as a young boy, he witnessed two cows separate from the herd and end up in the foyer of the Empire Theatre on James Street!

Wellington Dairy – from Duke Street, 2014. (Author)

Wellington Dairy – the yard and stables, Duke Street, 2014. (Author)

This new, larger business required a larger workforce. In 1899, Anthony placed two job adverts in the *Liverpool Mercury*; both emphasised milking ability:

Tuesday 1 August 1899 – Youth wanted, about 16 years of age, for milk house, must be a good milker. Apply Wellington Dairy, Garston.

Friday 22 December 1899 – Milkhouse – wanted, steady young man, 20–25, must be a good milker, good wages for a good man. Apply to Joy, Wellington Dairy, Garston, Liverpool.

At the beginning of 1901, the whole country was in mourning at the death of Queen Victoria. Nevertheless, in Garston, the silence of mourning was soon to be replaced by raised voices when the Corporation of Liverpool made representation to the Local Government Board for alteration of its boundaries to include the urban district of Garston. By then the district's population had grown to 17,288 and it had a housing stock of some 3,260 properties. An inquiry was held in June at Liverpool Town Hall and the cases for and against the proposal were hotly debated. The essence of this debate is described in the Reverend J. M. Swift's book, *The Story of Garston and its Church* (1937). Ultimately, the representation was granted and in the following year Garston was incorporated into the city of Liverpool.

1901 was also the year of the next census. The record shows that Anthony and his family were in residence at 37 Wellington Street: Anthony (age thirty-seven), Ann J. (age thirty-eight), Anthony P. (age eleven), George (age seven) and Flora (age four). Later that year, their fifth child, William Cecil, was born. He was christened at St Michael's, Garston Parish Church, on 15 October.

In 1903 Anthony indulged himself and his young family in a very special treat. The legendary western adventurer, Buffalo Bill, brought his Wild West Show to Liverpool. The show ran for three weeks at the Old Exhibition Ground on the Edge Hall Estate. As well as seeing the great man in the flesh the family were thrilled by a recreation of the famous battle with Sitting Bull.

On 23rd September 1905, Ann Jane gave birth to their sixth child. Kathleen Joy was christened at St Michael's, Garston Parish Church, on 5 December. With the birth of

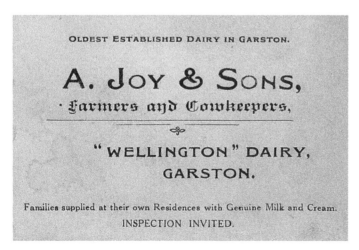

OLDEST ESTABLISHED DAIRY IN GARSTON.

A. JOY & SONS,

· Farmers and Cowkeepers,

◦

"WELLINGTON" DAIRY,
GARSTON.

Families supplied at their own Residences with Genuine Milk and Cream.
INSPECTION INVITED.

A. Joy & Sons business card. (Author)

Kathleen, the Joy household was now complete. Once his sons were old enough to help in the running the dairy business, they were acknowledged in the naming of that business. The family's business card proudly announced: A. Joy & Sons at Wellington Dairy.

By 1911 Wellington Dairy was flourishing. The Joy household consisted of Anthony (age forty-seven), Ann Jane (age forty-eight), Anthony Percival (age twenty-one), George (age seventeen), Flora (age fourteen), William Cecil (age eight) and Kathleen (age five). What the 1911 census does not show is that, as well as owning No. 37 Wellington Street, Anthony also owned the other three houses on the block: Nos 39–43. He was proving himself to be a keen businessman, and he was regarded as being firm, honest and someone who always expected and gave a square deal. He was a man of strong opinions but was ready to listen to others who saw things from a different angle. Any who chanced to disagree with him could not but admire his transparent sincerity.

Anthony was very active in the local community. He was a member of the Liverpool and District Cowkeepers Association for some forty years and was an official for many of those years. He was an Oddfellow for fifty years and reached high office in the order. Declining an offer to stand for election to the city council, he nevertheless took an active interest in local politics, and was a ward chairman for a number of years. In his later years he was president of the Garston Hotel Bowling Club.

He lived long enough to see all of his children happily married. Of his sons, it was only his eldest, Anthony Percival, who was to have children, or rather, a child – Anthony Eric Joy.

Anthony had a very special relationship with young Eric, the only one of his grandchildren to carry on the Joy name. Eric referred to Anthony as his 'papa' and the two spent many happy years working alongside each other at Wellington Dairy, – with Anthony helping to indoctrinate his grandson in the ways of the family business.

Anthony Joy and his grandson, Anthony Eric Joy, c. 1922. (Author)

On 22 February 1937 Anthony passed away and on 25 February was buried in the family grave. At his funeral there were many who came to pay their respects. His eulogy made reference to his Yorkshire roots and his family's cowkeeping business:

Anthony Joy was a well-known and highly respected figure in more than one sphere of local life. A Yorkshire man by birth he came to Garston when a young boy and was educated at our church day school. Here he has made his home for nigh 70 years. When a young man he joined his father in business and some here may remember the old dairy in Railway Street and the farm and the fields where now are the railway sidings.

When Anthony died, he left a wife, five children and three grandchildren. He also left Wellington Dairy in the capable hands of his two eldest sons, Percy and George.

The Cowkeepers of Garston

The Joys were not the only family to keep cows in Garston and they were not the only family with Yorkshire roots to do so. The property that Anthony Joy purchased in Duke Street had been used by a succession of cowkeepers who had also moved to Liverpool from the Dales. In 1881, Orlando Joy's brother-in-law, John William Daggett, was using the property. The Daggett family's occupation of 20 Duke Street, just across the road from the yard, lasted from approximately 1879 until 1891, after which they returned to Wharfedale to continue farming. By 1891 both the yard and the Dairy House were occupied by Joshua Burrow and his family, who were cowkeepers from Gargrave. The property became available when the Burrow family moved up Woolton Road to Willow Farm, on Long Lane.

The various trade directories published over the years include listings for 'cowkeepers' and for 'dairymen'; some businesses were listed under both headings. There is a third relevant heading – that of 'farmer'. There were a number of farms in and around Garston, indicative of its position on the edge of the built-up area. In its publication *Garston Farms and Past Inhabitants* (2000), the Garston & District Historical Society has included descriptions and records of the following farms:

- Willow Farm, Long Lane
- Dutch Farm, The Avenue
- Red House Farm, Mossley Hill Road
- Otterspool Farm, Jericho Lane
- Chapel House Farm, Bank's Lane
- White House Farm, Garston Old Road/St Mary's Road
- Beechwood Farm, Beechwood Road
- Island Farm, Island Road
- Holly Farm, Woolton Road

These were all traditional farms, having a main farmhouse (and collection of out-buildings) surrounded by fields. The cowhouses differed from the farms in that they were a part of the built-up area. If they did have access to fields for grazing, then

Cows being herded along Aigburth Road, Garston, 1950. (Frank Smallpage)

these were remote and necessitated a twice-daily trek from field to shippon for milking. There was at least one farm that had to adopt this cowhouse practice, once its fields and its farmhouse became separated by new building development. Right up until the mid-1900s, the cows of Wood End Farm (known as Dugdale's Farm) were grazing in fields off Aigburth Road but were herded from there to the farmhouse on Grassendale Road for milking.

Up until its incorporation into the city of Liverpool in 1902, Garston was included in the directories covering Lancashire. The table below includes names and addresses of cowkeepers and dairymen in Garston, as listed in the following directories: Kelly's 1895 (Lancashire), Gore's 1911 (Liverpool) and Kelly's/Gore's 1938 (Liverpool).

By cross-referencing these directory entries with census data it is possible to identify some of those families that had their origins in the Dales.

Property Address	Occupant (Directory Year)
37 Wellington Street	Joshua Burrows (1895) Anthony Joy & Sons (1911) (1938)
55 Window Lane	J. B. Cash (1895) Thomas H. Warth & Co. (1911)
1 Canterbury Street	Thomas England (1895) John Carr Davidson (1911) (1938)
23 Island Road	Anthony Joy (1895)
10 Railway Street	Daniel Joy (1895)
1 Russell Place	Robert Kirkley (1895) John Capstick (1911)
60 Chapel Road	Lawrence Mason (1895) (1911) Agnes Mason (1938)
4 Chapel Road	Charles Sutton (1895) Sutton & Kirby (1911)
25 Window Lane	Wm. Blackwell (1911) Mary Elizabeth Blackwell (1938)
2 McBride Street	Thos. J. Mason (1911) Mark Rhodes (1938)
3 Granville Road	Jn. Richardson (1911)
14 King Street	Robert Wilkinson (1911)
32 Lincoln Street	Warwick Birch (1938)
172 Garston Old Road	Cunningham & Cushen (1938)
194 Garston Old Road	Frank Harrison (1938)
1 Shakespeare Street	Wm. Edgar Heslop (1938)
10 Woolton Road	E.J. & O. Johnson (1938)
Fern Cottage, 33 Garston Old Road	Alfd. Rowbottom (1938)
195 St Mary's Road	Fredk. A. Shaw (1938)
41 Church Road	Mrs Clara Wrigley (1938)

Directory listings of cowkeepers and dairymen at Garston.

1 Canterbury Street

This end-terrace property had an access down the side that led to a shippon and yard at the rear. The access would have run behind properties on Window Lane. When these properties were demolished, the gable end of the cowhouse was exposed. Prior to its demolition in 2015, the rendering on the exposed gable end peeled away to reveal a ghost sign for 'Davidson's Dairy'.

1881 census –

Robert Blackburn (35) [b. Slaidburn, Yorks.] Cowkeeper/Dairyman. Mary Blackburn (34) [b. Long Preston, Yorks]. Children: Percy (3) [b. Childwall, Lancs.].

Robert Blackburn, born in 1846, was the son of William and Ellen Blackburn, who originally farmed at Swinshaw – a 184-acre farm in the parish of Easington that was lost when the Stocks Reservoir was constructed in the 1930s. By 1871 the family were farming 115 acres near Long Preston, where Robert was to meet his wife-to-be, Mary. Robert and his family were still living at Canterbury Street when, in 1886, Robert died prematurely at the age of forty years. He was buried at St Michael's Church, Garston.

1891 census –

Bryan Blackwell (51, Widower) [b. Gisburn, Yorks.] Cowkeeper. Children: Ann (22), Isabella (20), William (18) and Richard (16) [all born in Slaidburn, Yorks.].

Bryan Blackwell, born 1839, was the son of Bryan and Isabella Blackwell who farmed (The Clough and Higher Clough) in the parish of Easington. By 1871 Bryan had married Isabella and was farming 47 acres at Parrock Head, Slaidburn. Then, by 1881, he was farming 180 acres at New Close, Slaidburn. After running a dairy in Canterbury Street, Bryan returned to the parish of Easington to farm at The Clough in 1901, aided by his two brothers and sister. He died in January 1909 at Higher Clough.

Richard Blackwell, born 1876, was the youngest son of Bryan and Isabella Blackwell. Richard followed in his father's footsteps and, after marrying local girl Mary Elizabeth

1 Canterbury Street, Garston. (Author)

Bryan Blackwell – cowkeeper. (Andrew Mellin)

Barlow, became a dairyman at 3 Lodge Lane in Toxteth, Liverpool (1901). By 1911 he had returned to Garston and was keeping cows at 25 Window Lane. Mary Elizabeth continued to run the dairy after Richard died in 1933.

1901 Census –

John C. Davidson (23) [b. Bentham, Yorks.] Cowkeeper. Jessie Davidson (23) [b. Garston]. Children: William Henry (9 months)[b. Garston].

1911 Census –

John Carr Davidson (33) [b. Bentham, Yorks.] Cowkeeper & Dairyman. Jessie Davidson (32) [b. Garston]. Children: William Henry (11), Elizabeth Ann (9), Fred (5), Emily (3), Thomas (2) and Joshua (10 months) [all born in Garston].

John Carr Davidson – cowkeeper. (Carole and Clive Davidson)

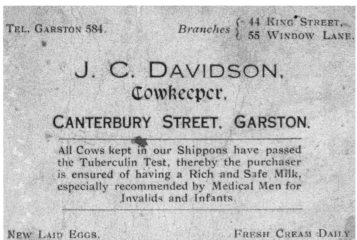

TEL. GARSTON 584.   *Branches* { 44 KING STREET,
                                   55 WINDOW LANE.

J. C. DAVIDSON,
Cowkeeper,

CANTERBURY STREET, GARSTON.

All Cows kept in our Shippons have passed
the Tuberculin Test, thereby the purchaser
is ensured of having a Rich and Safe Milk,
especially recommended by Medical Men for
Invalids and Infants.

NEW LAID EGGS.   FRESH CREAM DAILY.

J. C. Davidson business
card. (Barbara Price)

John Carr Davidson, born 1877, was the son of William and Ann Davidson (née Carr) who farmed at Summersgill, near Wray, on the border of Lancashire and Yorkshire. By 1891, the family had relocated to 1 Holland Street in Walton, Liverpool, to become cowkeepers. This is where John met his wife-to-be Jessie (née Moody), and they were married in 1898. The cows were milked in the shippon to the rear of 1 Canterbury Street but, according to Carole and Clive Davidson, the herd was grazed on Tushie's Field, Speke Road – land owned by the Tushingham family who farmed at Mersey View Farm. The Davidson business card shows that, as well as a cowhouse at Canterbury Street, the family also had dairy branches at 44 Kings Street and at 55 Window Lane. Indeed, John Carr Davidson was living at 55 Window Lane when he died at the age of sixty in 1944. John and Jessie had thirteen children, all born in Garston.

60 Chapel Road

The properties 60–64 Chapel Road form a terraced block on the corner of Chapel Road and Granville Road. The corner property was once a shop and there is a shippon to the rear of the block that can be accessed from the side entry on Granville Road.

1881 census –

John Uttley (28) [b. West Witton, Yorks.] Cowkeeper. Mary A. Uttley (34) [b. Leyburn, Yorks.] Edward Uttley (17), brother [b. West Witton, Yorks].

John Uttley, born 1853, was the son of Edward and Elizabeth Uttley. The family farmed in West Scrafton near Leyburn in Yorkshire, where John returned after he finished cowkeeping in Garston.

1891 census –

Thomas Lowes (32) [b. Muker, Yorks.] Cowkeeper. Dina I. Lowes (30), sister [b. Muker, Yorks.] Ralph Lowes (20), brother [b. Muker, Yorks.].

60–64 Chapel Road, Garston in 2015. (Author)

Rear yard and Shippon – 60–64 Chapel Road, Garston. (Author)

Thomas Lowes (also Lowis), born 1859, was the son of Ralph and Agnes Lowes, who farmed at Calverts House, Muker. After cowkeeping in Garston, the two brothers and one sister relocated to farm at Clivinger, near Burnley.

1901 census –

Lawrence Mason (46) [b. Cautley, Yorks.] Cowkeeper. Mary Mason (44) [b. Dowbiggin, Yorks.] Children: Thomas J (23), Richard J (16) and Charles (12) [all born in Dowbiggin/Cautley, Yorks.].

1911 census –

Lawrence Mason (56) [b. Sedbergh] Cowkeeper. Mary Mason (54) [b. Sedbergh, Yorks.] Children: Charles (22) [b. Sedbergh, Yorks.] Grandchildren: Lawrence (4) [b. Garston].

Lawrence Mason, born 1854, was the son of Thomas and Jane Mason, who farmed 25 acres at Hebblethwaite Hall, Sedbergh. In 1875 he married Mary Jackson and they had four

Cows at Garston Cross – corner of Chapel Road and Granville Road, *c.* 1910. (Garston & District Historical Society)

children. Their oldest son, Thomas James Mason, was born in 1878. By 1881 Lawrence was working as a butcher and he and his family were living with Mary's father at Wilkin Style, Sedbergh. Once he had become a cowkeeper, Lawrence did not return to farming in Yorkshire. When he died in 1934, he was living at Highlands, Pex Hill, Cronton.

3 Granville Road

Successive occupants of number 3 Granville Road were listed as being cowkeepers. To the rear of the property is a yard with a number of outbuildings – today accessed via entries to the sides of numbers 3 and 7 Granville Road. This yard and its associated buildings are good candidates for the former cowkeeping premises.

1891 census –

Henry Staveley (32) [b. Sedbergh, Yorks.] Cowkeeper. Isabella Staveley (33) [b. Kirkby Stephen, Westmorland]. Children: Nancy (5) [b. Liverpool].

3–7 Granville Road, Garston, 2015. (Author)

Rear yard and outbuildings at 3–7
Granville Road, Garston. (Author)

Henry Sedgewick Staveley, born 1859, was the son of David and Alice Staveley, who farmed 190 acres at Lunds, near Dent. The Staveley family farmed extensively in the Dentdale area. In 1881 he married Isabella Harrison before they relocated to Garston. By 1901, Henry had moved out of town to farm near Tarbock, where he lived until his death in 1920.

1901 census –

John Richardson (46) [b. Ravenstonedale, Westmorland] Cowkeeper. Agnes Richardson (44) [b. Sedbergh, Yorks.] Children: Maggie (16), Sarah A (13) and Isabella (9). [all born in Sedbergh, Yorks.].

John Richardson, born 1854, was the son of Thomas and Agnes Richardson, who farmed 21 acres at New House in Ravenstonedale, Westmorland. By 1891 he was farming at Low Riding near Sedbergh. After keeping cows at Granville Road, he moved round the corner and continued his cowkeeping at 64 Chapel Road (1911 census).

2 McBride Street

2 McBride Street, Garston. (Author)

Rear yard, now bricked up, 2 McBride Street, Garston (Author)

This is probably one of the best remaining examples of a purpose-built cowhouse in Garston (if not in Liverpool), situated on the end of a terraced block with a side entry leading to a yard and shippon at the rear. A feature of particular note is the cowkeepers' weather vane.

1901 census –

Alfred Jackson (34) [b. Bredbury, Cheshire] Farmer & Cowkeeper. Edith Jackson (34) [b. Helsby, Cheshire]. Children: Edward N (9), Gladys M (6), Alfred I (3) and Earnest (4 months) [all born in Garston].

1911 census –

Thomas James Mason (33) [b. Sedbergh, Yorks.] Farmer & Cowkeeper. Sarah Ann Mason (34) [b. Leeds, Yorks.] Children: Olive (7) [b. Garston].

Thomas James Mason, born 1878, was the son of Lawrence and Mary Mason who had kept cows at 60 Chapel Road. He married Sarah Ann Levitt in 1901 and by 1911 the couple were running the dairy at 2 McBride Street, where they were living with their daughter, Olive Mason.

Weather vane – 2 McBride Street, Garston (Author)

A date plaque, seen in 2016 at
2 McBride Street, Garston. (Author)

1 Russell Place

Russell Place was a terraced courtyard. Although it has since been demolished and replaced by a development of bungalows, according to local residents the building in the photograph is in fact the shippon that stood at the rear of 1 Russell Place. An original street sign for Russell Place is still attached to the side of the building.

1881 census –

John Walker (43) [b. Lofthouse, Yorks.] Cowkeeper & Dairyman. Eden Walker (43) [b. Lodge, Yorks.]. Children: Ann (18), Edith (15) and Ellen (9) [all born in Allershaw, Yorks.] 2 cowmen – Ireland Brows [b. Kirkby Malzard, Yorks.] and John Pickup [b. Skipton, Yorks.].

John Walker, born 1837, was the son of John and Eden Walker, who farmed at Thwaite House, Ramsgill in Nidderdale. In 1860 he married Eden Allen and farmed with his in-laws at the Lodge Farm Estate (250 acres) Stonebeck Up, near Middlesmoor. By 1871 he was farming 100 acres at Bramley Grange, Grewelthorpe. Once he had become a cowkeeper, John did not return to farming in Yorkshire. After leaving Russell Place he continued to keep cows at 15 Castle Street in Woolton (1891 census) and then lived with his cowkeeping

1 Russell Place, Garston. (Author)

son-in-law at 19 Breeze Hill, Walton (1901 census). By the time of the 1911 census, John and his wife were retired and living at 96 Elmdale Road in Fazakerley, Liverpool.

1891 census –

Robert Kirkley (39) [b. Nidderdale, Yorks.] Dairyman. Elizabeth (41) [b. Newbiggin, Yorks.]. Children: Margaret (13), Jane (11), Elizabeth (9), Joseph R (7) [all born in Nidderdale] and Deborah (5) [b. Garston].

1901 census –

Robert Kirkley (49) [b. Pateley Bridge, Yorks.] Cowkeeper & Dairyman. Elizabeth (51) [b. Pateley Bridge, Yorks.] Children: Jane D (21), Elizabeth (19), Joseph R (17) [all born in Pateley Bridge], Deborah (15) and Georgina (11) [both born in Garston].

Robert Kirkley, born 1852, spent the early part of his life living with his uncle, Joseph Kirkley, and his family, who farmed 200 acres at Low Woodale Farm in Stonebeck Up. He married Elizabeth Dent in 1876 and by 1881 was farming 420 acres at Newhouses Lodge Farm, Stonebeck Up. Once he had become a cowkeeper, Robert did not return to farming in Yorkshire. His wife died in 1909 and by 1911 Robert was living with his daughters at 52 Chapel Road, Garston and working as an insurance agent. He died in 1916.

1911 Census – James Herbert Preston (49) [b. Skipton, Yorks.] Dairy Manager. Dorothy Preston (40) [b. Settle, Yorks.] Children: Clifford (14), Rowland (12), Norman (10) and Dorothy (8) [all born in Seaforth/Litherland].

James Herbert Preston, born 1862, was the son of Robert and Ann Preston, who lived in Gargrave, Yorkshire. The family had no particular background in farming, although at one point James was a butcher. In 1901 he was married, still living in Gargrave and working as a bluestone quarryman. After living in Garston, James emigrated to New Zealand, where he died in 1916.

A. Joy & Sons

In 1914, Britain went to war. Percy Joy was by then twenty-four years old and, as cowkeeping was not a reserved occupation, he was called up and joined the King's Liverpool Regiment – Private 77254. When it came to duties and responsibilities in the army, Percy's background and working knowledge of farming and food production made him the obvious choice as designated cook.

The First World War affected the cowkeepers in a number of ways, perhaps the most obvious being that many of the menfolk enlisted and went away to fight. Businesses were maintained by older relatives coming over from the Dales and by the women taking on the duties previously carried out by their men. The war effort absorbed huge resources and cattle and fodder became scarce, forcing up prices.

Initially, the Liverpool & District Cowkeepers Association decided to 'put patriotism before profit' and, at their annual general meeting in February 1915, agreed to maintain the

Anthony Percival Joy, Private 77254, *c.* 1915. (Author)

price of milk at 4*d* per quart. However, with some cowkeepers being forced out of business, the association soon had to reconsider its position and a special meeting was convened in the July to review milk delivery in the city.

The Forage Department of the War Office had taken the best hay to supply the horses at the front, causing the price of what was being brought to market to be 50*s* per ton more than normal times. This, combined with the increasing cost and scarcity of cows, had greatly affected profitability – price was subsequently increased to 5*d* per quart. Also, shortage of labour meant it was difficult to maintain two deliveries per day; it was decided that, from 1 October until 30 April, milk would be delivered only once a day, between the hours of 6.00 a.m. and 12 noon, Sunday and weekdays.

By January 1916, the 353 members of the association had agreed to increase the price of milk again, to 6*d* per quart. Then, two years later – and under direction of the Food Control Committee of Liverpool – the price was fixed at 8*d* per quart, 4*d* per pint, 2*d* per half pint and 1*d* per gill. It was only in June of 1918 that the price increase was halted, returning to 7*d* per quart.

The loss of life during the war was immense and the city of Liverpool had its share of fighting men who were killed in their prime. Furthermore, due to its maritime situation, the city also lost many who were non-combatants. One incident in particular is worthy of note. On Thursday 7 May 1917, the Cunard-owned ocean liner *Lusitania* was torpedoed and sunk off the southern coast of Ireland. Of 1,959 people on board, only 761 survived. In their book *The Last Voyage of the Lusitania* (1956), Adolf and Mary Hoehling describe how in Liverpool the news of the sinking brought great anguish to hundreds of homes. For, although the passengers were mainly Americans travelling from New York to Britain and Europe, the crew were natives of the ship's home port – Liverpool.

Upon receiving news of the sinking, people took to the city streets in despair and large crowds gathered outside the Cunard offices, anxious for word about possible survivors. There was an international outcry and condemnation at the loss of innocent life. It is widely held that it was the sinking of the *Lusitania* that led to America entering the war.

Not all of the passengers were American, with the transatlantic services being frequented by British citizens who had family connections in America, including families who had been part of the mass exodus from the Dales. Percy had a second cousin, Hilda Mary Joy. She had married a Mr Norman Stones from her hometown of Ilkley and the couple had emigrated to Canada, where they were running a cattle ranch near Van Anda on Texada Island, Vancouver. Mr and Mrs Stones were returning to England aboard *Lusitania*. The news eventually reached Percy that, although Mr Stones had survived the sinking, Hilda Mary had not – her body was never recovered.

Percy served in France and at the end of the war was awarded the British War Medal. He managed to help defeat the Kaiser, survived the trenches and came home with all his limbs intact – only to return to his farming way of life and lose his thumb in a turnip-chopping machine. He used to say that the stump was real handy for tamping down the tobacco in his pipe.

While home on leave, Percy married Ellen Savage of 1 Railway Street. The ceremony took place at St Michael's, Garston Parish Church on 24 July 1916. After the war, Percy and Ellen moved into 39 Wellington Street, next door to the Dairy House. This was an ideal arrangement as it meant that Percy was on site to continue working in the family business alongside his father, Anthony, and his brother, George.

Anthony Percival Joy, *c.* 1916. (Author)

Ellen Savage, 1916. (Author)

On 18 December 1919, Ellen gave birth to their first and only child, Anthony Eric Joy. He was christened Anthony in what by then had become the family tradition of naming the first-born son, and Eric, after Percy's cousin, George Eric Wright, who had been killed in action. Percy wasted no time in introducing young Eric into the farming way of life. As soon as the boy could walk he was accompanying the adult members of the family as they went about their work.

Eric's earliest memory was when he was just two years old, sitting in the milk float on his father's lap, going under the bridge at Allerton station. His father was holding him with one hand and driving the horse, Daisy, with the other. By the time he was four years old Eric was helping out in all aspects of the job – and he had the scars to prove it. One day when he was feeding Daisy, he didn't hold his hand quite flat enough and, before he knew what had happened, she had taken his fingernail off. Six months later he was leading Daisy down The Avenue, going across the bridge, when a train came underneath her. She started to dance and trod on Eric's foot; he lost his big toenail. Nevertheless, he absolutely loved

Above: Young Eric with his dog at the entrance to the yard, Duke Street, *c.* 1923. (Author)

Left: William Cecil Joy and young Eric, *c.* 1924. (Author)

Mucking out – Eric in the yard at Wellington
Dairy, *c.* 1930. (Author)

Haymaking – Eric helping his father in the fields,
c. 1927. (Author)

every aspect of working with his family and was not at all happy when, in 1925, he had to leave his daily routines of delivering milk or working in the fields, in order to go to school.

In 1926, George married Mary Westwood, a widow from Northern Ireland. Her husband, who had been stationed over in Ireland with the Royal Ulster Fusiliers, was killed in action. So, Mary came over to Garston to live with her husband's family in Clifton Street and that is how she met George. Initially, they moved into rented accommodation over Chapman's shop in St Mary's Road.

By 1931 the Great Depression had begun. People were often to be seen begging in the streets or, worse still, fainting through lack of food. One chap came knocking at the Dairy House, seeking work. He had walked and hitchhiked from Hull in search of employment. When he was told that there was no work for him, he decided to walk back to Hull. Percy invited him in and gave him a cup of tea and a sandwich to help him on his way. On another occasion Percy was out in the milk float with Eric when a man fell flat on his face in the middle of the road. Percy pulled over and bought the man a cup of tea and a pasty from the nearest cake shop. At a time when people were starving, the Joy family were sustained by Wellington Dairy – and they were grateful.

By the time Anthony Joy passed away, in 1937, Percy and his family were living at 37 Wellington Street and George and Mary were living next door at 39 Wellington Street. The youngest of the Joy brothers, William Cecil, had married and was running a butcher shop at 87 St Mary's Road. The sign in the window read 'Have you ever had the Joy of enjoying Joy's sausages?' Following the death of their father, Percy and George continued the business and retained the family name – Anthony Joy & Sons.

The brothers prepared to face a new era together. They had a dairy herd of some twenty-four cows, two horses for delivering the milk and the family also kept pigs. They had a shed on The Avenue, which they used as a loose box for calving cows and also as a boiler house for boiling the swill for the pigs – nine breeding sows and one boar. The local chip shop would donate their used peelings and kids would bring their family leftovers in exchange for a few sweets. It was all boiled in a barrel and fed to the pigs. They would get two litters a year and would sell the young 'suckers' at ten weeks old to the sewerage farms at Fazakerley, which were run by Liverpool Corporation.

WILLIAM JOY,
BUTCHER,
TRAVELLING SHOP.
—/—/—
ONE QUALITY —— THE BEST.
—/—/—
CUSTOMERS VISITED DAILY
: AT THEIR HOMES. :
87 ST. MARY'S ROAD, GARSTON

W. C. Joy – butcher. Advert in parish magazine. (Author)

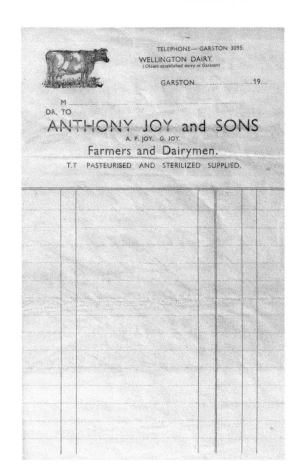

Right: Anthony Joy & Sons headed business stationery. (Author)

Below: Grazing cattle – The Avenue, Garston, 1930. (Author)

George Joy with 'Daisy' at The Avenue, Garston, 1921. (Author)

Two of their horses were Sam and Bonnie. Sam was a 13.2-hands-high pony who was purchased in 1923 at the age of twenty-seven and remained with the family until 1942, when his kidneys failed. Apparently, Sam was quite a character and used to bite the buttons off your waistcoat when you were grooming him. Bonnie arrived in 1926 and worked at Wellington Dairy for fifteen years until 1941, when she died of pneumonia. Eric also had a goat called Nancy. She was a gift from Mr Mullet, the saddler in Garston. He kept two goats (at Mersey View) because his daughter used to have sleeping sickness; it was said that goat's milk was naturally tuberculin-tested, as goats did not get TB. One of the customers made a cart for Nancy and Eric used to drive her around the streets, pulling the cart.

Things were continuing to change in Garston and the expanding population had need of more land on which to build houses, schools and shops. Consequently, in 1938, the brothers were served notice to quit their farming of the fields along The Avenue. Percy and George were able to rent from the church one small field (Joy's Field, off the newly built Horrocks Avenue) that was just big enough to give an annual crop of hay. But, without fields for grazing, the milking herd had to be relocated to the shippon at Wellington Dairy. In an ironic twist, the Joy brothers had to revert back to the cowkeeping practices of their inner city compatriots.

'Bonnie', waiting patiently for the cart to be repaired. Woolton Road, Garston, c. 1940. (Author)

Country Cousins

Like most of the Dales folk who relocated to Liverpool, the Joy family maintained contact with their relatives back in the Dales. Indeed, it was said of Anthony Joy that he was as well known in the Craven district as he was in Garston. And no doubt his presence there was as much to do with business, as it was to do with visiting cousins. Although Anthony had a number of cousins, there was one branch of the family with which the Joys were particularly close – the Hardacre and Metcalfe families. The Joys had become related to the Hardacres and the Metcalfes through marriage:

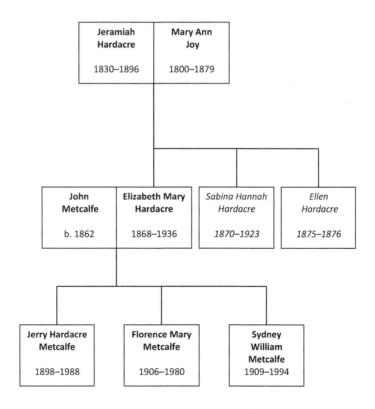

Daniel Joy's older sister, Mary Ann Joy, married Jeramiah Hardacre. They had three daughters, Elizabeth Mary, Sabina Hannah and Ellen (who died as an infant). These three girls were all cousins of Anthony Joy. The eldest of the three, Elizabeth Mary Hardacre, married John Metcalfe. Their three children were Jerry, Florence and Sydney Metcalfe – second cousins of Percy Joy. The family lived in Threshfield where John Metcalfe was a carter working for a local coal merchant.

By the time Percy was the head of the family it is less likely that he travelled to Yorkshire for business purposes, as all of his business needs could by then be met locally in Liverpool. When Percy took his family to visit his Yorkshire relatives, it was also to enjoy a holiday in the countryside.

In 1926 Percy and his family spent ten days in Grassington, staying with a Miss Chester. Percy would take a maximum of twelve days' holiday at a time so that he was only away for one weekend in the fortnight. He always started his holiday on a Monday and returned a week the following Friday so that he was back for the weekend, in time to collect the milk money – a very important task.

This holiday was particularly memorable because of the difficulties in travelling – the General Strike was on and the railways were running slow. A journey that would normally have taken four hours by train, from Liverpool to Threshfield, ended up taking ten hours. They had to change at Preston, then Blackburn and then Colne, where they waited for two hours. From there they went on to Skipton and then caught the local train to Threshfield. They set off at nine o'clock in the morning and arrived at Grassington at seven o'clock in the evening. But the long journey did not spoil an otherwise thoroughly enjoyable holiday.

In the summer of 1928 Percy took his family back to Yorkshire, this time staying in Threshfield. On this occasion Percy's mother, Ann Jane, accompanied them. Unfortunately, while they were away, Percy took ill. He was so ill that the doctor had to be sent for in the middle of the night. The doctor kept him in bed for three days while Ellen dosed him with whiskey; apparently, that treatment eventually brought his temperature down. He spent the rest of that holiday just walking around Threshfield, getting back on his feet again.

These strong family ties were also maintained through a reciprocal arrangement, whereby the Joy family played host to their Yorkshire cousins who came to the city to

Percy with Ellen and sister-in-law Annie [left], on holiday beside the River Wharfe, c. 1926. (Author)

enjoy a working holiday. There was a particularly strong bond with Florence Metcalfe, who visited and stayed at Wellington Dairy on more than one occasion. Indeed, the closeness of the family bond was demonstrated when Florence was bridesmaid at Kathleen Joy's wedding at St Michael's Church, Garston, in 1931.

Eric, Percy and Ann Jane on holiday in Threshfield, 1928. (Author)

Anthony Joy's cousins – Elizabeth Mary Metcalfe and Anthony Joy (of Hole Bottom), c. 1915. (Author)

Left: Percy Joy's second cousin, Florence Mary Metcalfe, at Wellington Dairy in 1922. (Author)

Below: Jeremiah Metcalfe, Percy Joy, Eric Joy (baby), Florence Metcalfe, George Joy and Ivy Branegan. Wellington Dairy, *c*. 1920. (Author)

Cowkeeper Turns Milkman

Eric had what he considered to be an idyllic childhood. He spent most of his time working with his family, either in the fields, in the yard or out on the milk rounds. Each summer was punctuated with a family holiday. Not all such holidays were spent revisiting the place from which the family had sprung – the Yorkshire Dales. Eric also fondly remembered holidays in Blackpool, Grange-over-Sands and Penmacno.

However, this childhood idyll was snatched away in the starkest of ways. On 3 September 1939, two days after Germany invaded Poland, Britain and France declared war on Germany, and on 3 April 1940, at the age of nineteen, Eric was conscripted into the army. The cowkeepers were to go to war, again.

Eric served in the Royal Army Veterinary Corp. – Private 3534215. There he found himself serving with people of the same ilk to himself: farmers, jockeys, grooms, stable workers and gypsies – people who spent their lives working with animals on the land. There he forged friendships that were to last a lifetime. Among his closest friends were: Bernard Vanstone, a Devonshire farmer; Wilf Boulton, of Limestone Stud in Lincolnshire; George Pemberton, a saddler from Yorkshire; and, champion jockey and fellow Liverpudlian, Joe Sime.

Although Eric was trained as a telephonist, he spent most of his time working on farms in Lancashire. Due to the demands of the war effort, soldiers were assigned to work on farms to maximise domestic food production. So, there was Eric, serving his time in the army working on farms in Lancashire – Veterinary Corp. soldiers were in big demand by local farmers. Then, as soon as he was on leave, he would be straight back to Garston working on the milk rounds. He was often to be seen delivering milk dressed in his army uniform. He also found time to be best man at his cousin Wilfred Percival's wedding in 1943.

Eric's army book contains a complete record of the leave he was granted during his military service (p. 99). As well as registering the periods of leave to which Eric was entitled (referred to as 'Privilege' leave), the record includes periods of leave described as 'Agricultural'. This agricultural leave was granted to military personnel who were not exempt from service, but whose non-military skills were pertinent to maintaining the war effort. In Eric's case, he would have been granted agricultural leave to return home and work at the family dairy. Over 40 per cent of his total Veterinary Corp. leave was granted as agricultural leave.

Eric Joy – Private 3534215, Royal Army Veterinary Corp., 1944. (Author)

Eric was discharged from the army at Melton Mowbray on 31 May 1946. His military conduct was described as 'Exemplary' and he received the following testimonial:

A hardworking and conscientious man. Has a good knowledge of horses and stable work.

Throughout the wartime period Wellington Dairy was supplied with brewery grain, used as a dietary supplement for the cattle. Brewers' grain was the solid residue left after the processing of cereal grains, mainly barley, for the production of beer. In the brewing process, grains were soaked in water until they germinated and were then dried to produce the malt. This malted grain was then milled and steeped in hot water so that enzymes could transform the starch into sugars. Once the liquid was removed to produce beer, the remaining 'grain' was a concentrate of proteins and fibre that was suitable as an animal feed, particularly for ruminants.

Home on leave. Eric, best man at Wilfred Percival's wedding, with bridesmaid, Joyce Bridge, 1943. (Joyce Johnson)

Record of Eric Joy's Military (Agricultural) Leave				
The Royal Army Veterinary Corp. 1941–1946				
From	**To**	**Duration (days)**	**Type**	**Travel Warrant Issued**
10.10.41.	16.10.41.	7	Privilege	Yes
17.02.42.	24.02.42.	7	Privilege	Yes
21.03.42.	**17.04.42.**	**28**	**Agricultural**	**Yes**
15.05.42.	22.05.42.	7	Privilege	Yes
21.08.42	28.08.42.	7	Privilege	Yes
01.12.42.	09.12.42.	9	Privilege	No
03.03.43.	12.03.43.	9	Privilege	No
08.05.43.	10.05.43.	2	Privilege	No
04.06.43.	11.06.43.	7	Privilege	Yes
10.09.43.	19.09.43.	9	Privilege	Yes
21.11.43.	28.11.43.	9	Privilege	Yes
15.12.43.	**29.12.43.**	**14**	**Agricultural**	**Yes**
05.03.44.	14.03.44.	9	Privilege	Yes
15.09.44.	24.09.44.	9	Privilege	Yes
31.01.45.	09.01.45.	9	Privilege	Yes
04.05.45.	13.05.45.	9	Privilege	Yes
20.06.45.	**10.07.45.**	**21**	**Agricultural**	**Yes**
05.09.45.	17.09.45.	12	Privilege + 2VS	Yes
06.12.45.	16.12.45.	9	Privilege	Yes
19.02.46.	26.02.46.	7	Compassionate	Yes
27.02.46.	05.03.46.	7	Ext. Compassionate	No
06.03.46.	**19.03.46.**	**14**	**Ext. Agricultural**	**No**
28.03.46.	**25.04.46.**	**28**	**Agricultural**	**Yes**

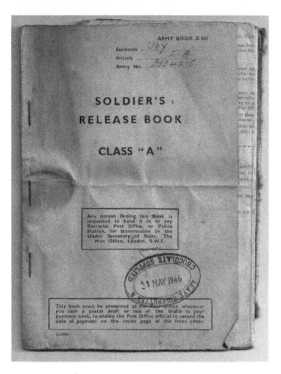

Eric Joy's army service release book, 1946. (Author)

Eric Joy's army service release certificate, 1946. (Author)

The grain was delivered via horse and cart by the Ranson family, who were cowkeepers based in Raffle Street, off Great George Street. They bought the grain from the city centre breweries such as Threlfall's in Truman Street and collected it using a sided cart. The cart was placed below a chute, down which the still hot and wet grain was dropped. By the time they arrived at the dairies, the grain had cooled and most of the moisture had drained away. Once they had reached their destination, the grain had to be shovelled off by hand and was sold to the dairyman by the bushel; one spadeful equalled one bushel. Arthur Jones related an incident that took place when his father was delivering grain to Wellington Dairy: 'My father had to back his wagon into the dairy yard to deliver the brewers' grain [and] as he got to the rear of the wagon it rolled back against the midden, injuring him.' Fortunately, the injury was not fatal – though, once recovered, Mr Jones Snr found alternative employment working at the Dunlop tyre factory in Speke.

During the whole of the war, there were only three land mines dropped on Garston by German aircraft, two of which did not explode. The city of Liverpool did not fare so well. Because of its prominence as a port, receiving supplies from across the Atlantic, the Luftwaffe selected it for special treatment. For eight successive nights from 1 May 1941, bombs rained down on Liverpool, killing 1,746 and injuring 1,154 others. More than 90,000 homes were destroyed or damaged and 75,000 people were left homeless. Many of the cowhouses were bombed out of existence. In his 1970 essay 'The Story of the Liverpool Cowkeeper', the dairy husbandry adviser for Lancashire, John Sumner, suggests that some of the bigger shippons may have been targeted as, from the air, they may have been mistaken for factories or warehouses. By 1949, only 129 cowkeepers remained.

After the war there were major changes in the way that milk was produced and distributed. Pasteurisation and refrigeration were more commonplace and this had helped generate a shift away from rail to road transport – increasing competition with the city cowkeepers. The table below illustrates this shift in the source of milk being supplied to Liverpool during the period 1927–1951.

In addition to these changes in production and distribution, a new legal framework was introduced that enforced the latest standards of health and hygiene. It became increasingly difficult for the small dairies to meet the costs required to comply with the new legislation – but not so for the rapidly expanding corporate dairies. Some of the Liverpool cowkeepers had literally been bombed out of existence during the war, but those who survived faced stiff competition from the corporate dairies. They had a choice between either selling up (and maybe returning to the Dales) or throwing in their lot with the corporate dairies and becoming the corporates' mechanism for delivering milk to the doorstep.

There was a hard core of cowkeepers in Liverpool who held out as long as they could in the face of these changes. The Joys were among this hard core and managed to survive

The Milk Supply of Liverpool 1927–1951 (Five year mean percentage)			
[Adapted from Grundy, 1982]			
Year	City Cowkeepers	Rail	Road
1927–31	27.1	27.3	45.6
1932–36	22.1	11.9	65.8
1937–41	16.8	7.7	75.5
1942–46	10.6	11.3	77.8
1947–51	6.3	1.1	92.7

Best milker in the herd – Wellington Dairy, Duke Street, 1952. (Author)

Liverpool & District Cowkeepers Association Annual Dinner, February 1952. Eric is on the fifth table from the front, far left. (Gilbert and Margaret Rowlands)

through compromise. Although they continued to produce their own milk, they sold it in its raw state to one or more of the corporate dairies. In a reciprocal arrangement, milk that had been treated, chilled and bottled was then bought back by the dairy.

By this time, the war between the cowkeeper and the corporate dairies had run its course. Indeed, the two suppliers of milk used by Wellington Dairy both had a close empathy with the cowkeeper that was. The first, Dairy Farmers Creamery (known as D.F.C.), was a co-operative in which many of the small dairymen and farmers in and around Liverpool had shares. The second supplier was J. Hanson & Sons. James Hanson had himself been a farmer who had become a cowkeeper and dairyman before selling his herd and establishing a corporate dairy of his own.

Both of these companies delivered crates of cooled, treated, and bottled milk to Wellington Dairy. The milk lorries would drive up Duke Street and pull up at the yard entrance. The boiler-suited deliverymen would then unload a crate at a time and carry it into a walk-in refrigeration unit that had been installed in the lean-to. The milk was delivered to the dairy in the afternoon, stored overnight in the refrigerator and then delivered to the customer's doorstep the following morning.

Such major changes in the milk industry necessitated the redrawing and reallocation of the milk rounds. Wellington Dairy was allocated two rounds. The first, the early morning round, was 'under the bridge' – that is, under Church Road Bridge. We called this 'the bottom round' and it included Windfield Road and its adjacent streets. The second, the mid-morning round, was 'over the bridge' – that is, up and over Chapel Road Bridge. We called this 'the top round' and it included the streets off Duncombe Road South, between Garston Old Road and St Mary's Road.

The bottom round was a two-man job and, once Percy began to feel his age, it was left to Eric and George. For this round they used the covered van, with the milk crates stacked at the rear. Neither milkman ever used a written list of customers. They both knew from memory who lived where and what their daily order was. If a customer wanted to change their order on a particular day, they left a note in one of their empties. So, they always carried slightly more milk than had been ordered. The milk came in two sorts: sterilised and pasteurised. Bottles of 'sterrie' came with long necks and a metal cap, which you could only remove with a bottle opener. Bottles of pasteurised were your common milk bottles with the silver foil tops; they came in pint and half-pint sizes. The two milkmen also carried pints and half-pints of orange juice.

Eric and George would load up their respective hand crates and start on the opposite side of the road from each other. Bottles would be delivered on foot to each doorstep and the empties collected. When the hand crate was full of empties they would return to the van, transfer their empties into the crates in the back of the van and top up their hand crate again with the right mix of bottles to serve the next set of houses.

However, this was a three-man team and the horse, Rupert, was the third man. Together, they worked the round in perfect synchronisation so that Eric and George did not need to drive the horse while they were delivering. As they worked their way along the two rows of houses, they would give the horse a call – 'Come on!' – and the horse would walk the van along the road, catch up with them and then stop. The horse would do this with no one at the reins and could even manoeuvre the van around parked vehicles. The horse's favourite stopping place was the 'green' in Monkfield Way. While Eric and George delivered to the surrounding houses, the horse would bump the van up onto the green, haul it to where the best grass was growing, and enjoy a mid-morning graze.

On one occasion, while doing the bottom round with Rupert, Eric had called the horse on. But after taking one step, Rupert stopped. Eric called him again, but Rupert just stood in the middle of the road and shook his head. When Eric went back to get him he found that a cleat had broken and one of the traces was trailing on the ground. Rupert had either felt or heard the difference and knew there was something wrong – he refused to budge until the problem was rectified.

The 'top' round was a shorter round and could be done using the smaller and lighter milk float. It was only possible to do the round with one person, usually Eric, because of the contribution of the horse – adopting a similar partnership approach as that used on the bottom round. Eric would deliver the milk on foot, to-ing and fro-ing between the two lines of terraced houses, while the horse, following on behind, would take a perfect line up the middle of the street. They also had a handful of customers who lived near to the dairy, off St Mary's Road. Milk was delivered to these local customers using a delivery bike with a milk crate mounted front and rear.

Eric continued the Joy family's long-standing association with St Michael's Church. Indeed, it was through the church's youth fellowship that he met his wife-to-be. Alice Atherton was born on 30 September 1924 and was one of eleven children of John Thomas

Traditional milk float pulled by Rupert, Garston Old Road, 1963. (Author)

Atherton (1896–1970) and Alice Ann Wilkinson (1899–1981). Eric and Alice were married on 31 July 1954 at St Michael's Church, Garston.

Once his sister became a member of the Joy family, Alice's younger brother, Ronald, spent time working at Wellington Dairy. This was to give Eric a hand on those occasions when George was taking a holiday. As well as being shown how to milk a cow, Ron's duties also included mucking out and feeding. The hay was obtained from a field in Speke. Ron and Eric would take a horse and cart out to the field in the morning and turn the cut hay with rakes to make sure it was dry. Later in the day a chap from Halewood would turn up with a bailing machine. Once the cart was loaded with wire-bound bales, the hay would be taken back to the yard to be stored in the hayloft.

Ron recalls working at the dairy on two occasions – the first in 1954, after Alice and Eric were married, and the second in the summer of 1955. On both occasions cows were being kept in the shippon in Duke Street. However, by the end of 1955, the cows were gone.

The decision to cease producing milk would have been a difficult one, both financially and emotionally. Financially, because without milk to sell to the corporate dairies, the profitability of the business was further reduced. Emotionally because keeping cows had been something that Percy, George and Eric had done all of their lives. Nevertheless, the Joys were no longer 'cowkeepers'.

Eric and Alice (far left) with members of the extended Joy family, attending the Liverpool Cowkeepers Annual Dinner at the Exchange Hotel, Liverpool, 1954. (Barbara Baker)

Liverpool – Last Stronghold of Town Cowkeepers

In his article, 'Liverpool – The Last Stronghold of the Town Cowkeepers', H. Hill, the county milk production officer for Lancashire, describes the cowkeeping situation in Liverpool in 1954 and reflects on the decline in the practice of cowkeeping over the preceding twenty years.

In 1934, Liverpool had 232 registered cowkeepers keeping approximately 3,500 to 4,000 cows and producing up to 10,000 gallons of milk per day. When the Milk and Dairies Regulations were introduced in October 1949, the number of registered producers had fallen to 107 and keeping 1,500–1,700 cows yielding 5,000 gallons of milk per day. By the end of 1954, the number of producers had fallen to just forty-three, with only thirty of these located within the built-up area proper. At that time it is probable that the Joy's business at Wellington Dairy would have been one of these remaining forty-three producers, although it is unlikely that it would have been considered part of the built-up area of the city.

The county milk production officer goes on to suggest that this decline in the number of cowkeepers was due to a number of contributing factors:

- The high cost of feedingstuffs – during and immediately after the war, the unique position of the town cowkeepers as primary producers was recognised by the allowance of an increased concentrate ration. However, once rationing ceased, the price of feedingstuffs became subject to market forces and escalated accordingly. Those cowkeepers who had access to pasture and/or meadow were able to offset the increases in cost, but those within the built-up area had to pay the prevailing market prices.
- The seven-day week – this is a reference to the fact that milk production is a seven-days-per-week job, whereas in other occupations a five-day-week had become the accepted norm. When combined with the lower rates of pay offered by the milk business in comparison to other occupations, it became more and more difficult for the cowkeepers to recruit suitable labour. This made the option of becoming retail dairymen (as opposed to producer-retailers) more attractive, as it eliminated the costs of feeding and caring for the livestock, the cost of milking and the costs of treating and bottling the milk. The retailing regime had a better fit with the five-day week.

- The Specified Areas scheme – under this scheme no non-designated milk could be sold by retail, the only grades of raw milk being available to the public being 'Tuberculin-tested' and 'Accredited'. Liverpool was one of the first provincial districts to be dealt with in this way, becoming a specified area in November 1952. The designated area included a number of districts abutting the city boundaries, but the threat had been there from the Ministry of Food for the whole of the city to be designated. Though some city cowkeepers did acquire Accredited or Tuberculin-tested licences, the threat of designation was enough to see many others give up milk production and become retailers. An additional direct outcome of the scheme was the inauguration of a co-operative dairy in Kirkby, just outside the city boundary. Those producers in the city, who had an interest in the venture, were able to send their raw milk to the co-operative for heat treatment, treated supplies then being returned for retail sale.

It was speculated that a further contributor to the decline of the cowkeepers in Liverpool may have been the level of control that was exerted upon them with regard to their premises. Across the country, control was originally afforded by the Milk and Dairies Order, 1926, and then by the Milk and Dairies Regulations, 1949. However, in Liverpool, this legislation was augmented by the Liverpool Corporation Act, 1921, continuing the long-term practice of dual levels of supervision within the city. Under the Corporation Act, premises for the keeping of cattle had to be licenced for this purpose by Liverpool Corporation and would be subject to inspection by officials of the Corporation; in a typical year, the Corporation could carry out in excess of 4,000 inspections. Also, the keeping of cows on unlicensed premises was subject to penalty. The permitted number of animals for each building had to be displayed by notice affixed to the main entrance to the said building. It was also a requirement of the Act that licensed premises were to be kept in good order at all times with regard to cleanliness, ventilation, water supply and drainage. Failure to ensure the required standards would result in licences being withdrawn. Although the strictness of this regime may have contributed to some cowkeepers giving up their dairy herds, it was noted that those who remained maintained their premises to the highest standards.

The Liverpool Corporation Act of 1921 required shippons to be licensed. (Author)

The article praised the condition of the Liverpool dairy herds. The Liverpool cowkeeper was credited with an understanding of how to feed his animals to maintain their condition and with a view to obtaining the maximum amount of milk, despite the large quantities of quality hay and lucerne that were imported from Canada prior to the war no longer being available. The milking animals of the city compared favourably with those kept under what were generally considered to be more normal conditions. Furthermore, not only were the animals and premises in excellent condition, but also the utensils were invariably kept clean. Steam sterilisation took place at least daily, which 'in these days of almost universal hypochlorite treatment, speaks much for the orthodoxy of these producers'. It was stated that, in view of the good methods employed, it naturally followed that the keeping quality of the milk was universally satisfactory, so that customers would have had little cause for complaint. 'A pride in doing a good job is general among the City's cowkeepers, and no efforts to achieve this end are spared.'

Mr Hill concludes his article by speculating on the future of the Liverpool cowkeeper. On the one hand, as 50 per cent of the cowkeepers at that time did not possess either Accredited or Tuberculin-tested licences, it was anticipated that numbers would drop once the city came within a scheduled Attested area. On the other hand, it was felt that those producers who had obtained Attested herds would continue and, indeed, their numbers might increase. On that basis it was postulated that Liverpool would remain the last stronghold of the town cowkeeper for many years to come. As it turned out, the Liverpool cowkeeper lasted another twenty years.

Liverpool's successive occupants of the position 'Medical Officer For Health' produced annual reports on the health of the city. Each of these reports included a section on the licensing of premises for keeping cows. The numbers of licences issued provide a guide as to the rise and fall of cowkeeping in the city over a period of some 100 years.

'A pride in doing a good job is general amongst the City's cowkeepers' (Frank Smallpage)

Number of Premises Licensed for Keeping Cows in Liverpool 1868–1974 (five-year means)

[Adapted from Grundy, 1982]

Date	Licences Issued	Year	Licences Issued
1868–70*	448	1921–25	292
1871–75	521	1926–30	278
1876–80	436	1931–35	265
1881–85	449	1936–40	210
1886–90	419	1941–45	158
1891–95	338	1946–50	119
1896–1900	433	1951–55	80
1901–05	447	1956–60	35
1906–10	467	1961–65	21
1911–15	427	1966–70	12
1916–20	347	1971–74**	2

*three-year mean. **four-year mean.*

One Hundred Years
of Delivering Milk

To begin with, Eric and Alice lived at 37 Wellington Street, sharing the two-up, two-down terraced house with Percy and Ellen. George was living next-door at 39 Wellington Street and, between them, Percy, George and Eric kept the business running. In 1956, Eric purchased his own house in Garston Old Road, overlooking Garston Park. He also bought himself a bike so that he could cycle the short journey to and from work, up and down Chapel Road.

Although the family had given up its milking herd, it had retained its use of the field off Horrocks Avenue. No longer required for grazing cattle, the field was turned over to hay. The pasture had become a meadow and the hay crop was used to feed the horses. However, in 1958, the church served the family with notice to quit its use of the field. The site was needed for the construction of a new school (Garston C of E Primary School) and vicarage. This act severed the last link between the family and working the land – the Joys were no longer 'farmers'.

The last hay crop – Joy's field, 1958 (George above and Eric below). (Author)

Eric and Alice went on to have three children and the family tradition of christening the eldest boy 'Anthony' was continued into this next generation. We three kids had a great time being a part of Wellington Dairy. We all have fond memories of accompanying our elders, going about their daily work with the three milk horses Danny, Rupert and Peggy – all Irish Vanners.

Garston Docks had a quayside timber mill and we would take the van down Dock Road and load it up with huge sacks of sawdust to be used for cleaning out the stables. Then, during the summer, Liverpool Corporation would mow the grass on Allerton Cemetery and leave it to lie. So, with their permission, we would take the van up there and spend the afternoon raking in as much grass as we could. While we were raking it in, Rupert would wander around the field pulling the van behind him, grazing away on the cut grass. Once we had a big enough stack we would load it onto the van until the sides were overflowing and then take it back to the stables for Danny and Peggy to have their share. Although we were taking the grass cuttings to be a dietary supplement for the horses, in doing so we were mirroring the practice of the early Liverpool cowkeepers, who took grass cuttings from the city's parks, gardens and cemeteries to feed to their cows.

One of our favourite trips was up to Blackmore's smithy in Woolton to have the horses shod. Bernard Blackmore was a hot-shoe farrier. His forge stood in an open-sided lean-to shed. Despite the shed being open to the elements, the heat of the forge filled the space beneath the rusty corrugated tin roof. You simply had to cross the threshold of the shed to walk into instant warmth that smelled of coal with a hint of metal. On the back wall of the shed was every possible size and style of metal horseshoe, at least four of each.

The Joys' milk horses – Percy with Danny, and George with Rupert and Peggy, c. 1966. (Author)

Eric and his three children with Danny, *c.* 1966. (Author)

It was fascinating to watch the 'smith doing the shoeing: using a buffer and hammer, he would straighten the old nails, or clinches, which had been bent over to hold the shoe in place; the old shoe would be loosened using a pair of pincers, the old nails pulled with the claw of the hammer and then the shoe removed; the wall of the hoof would be trimmed using a pair of nippers, and then the sole and frog of the hoof would be trimmed using a drawing knife. A shoe would be selected, tried for size and then buried under the red-hot coals of the forge until it softened.

Once the shoe was glowing red hot the blacksmith would thrust a pritchel through the last of its seven nail holes and lift it out of the forge; taking the horse's hoof between his chap-protected thighs, the blacksmith would place the red hot shoe onto the newly trimmed surface of the hoof; the shoe would hiss, spit and smoke as it burned into the hoof, giving off the most curious smell of something cooking!

Having burnt its mark on the hoof, the shoe was returned to the forge to reheat; using the burn mark as a guide, once the shoe was red hot again, the blacksmith would work it into shape on his anvil, using a hammer and tongs. The shoe was then returned to the forge one last time before being used to give a second burn on the hoof, just to make sure of a snug fit; the shoe was then cooled in a water tub before being nailed into place on the hoof (the nails were specially shaped so that they would turn out as they went through the hoof). Once the nails were through the hoof their sharp protruding points were bent over using a clincher and everything was smoothed off using a rasp. The final touch was to give the hoof wall a quick coat of oil with an old paintbrush. This process was repeated for each shoe that needed to be replaced.

While in Woolton, we would sometimes call in at the veterinary surgery – another supplier of services to the family business. Our vet was G. R. Sumner, who had a surgery at 58 High Street. Mr Sumner would occasionally come out to Wellington Dairy to see one of the horses, though these visits became more frequent as the horses grew older – especially so, with Danny.

A family tradition, handed down through the generations from the days of the Liverpool cowkeepers, was attending agricultural shows. The two main local shows were the Liverpool Show, held at Wavertree Playground (The Mystery), and Woolton Show, held at Camp Hill in Woolton. Occasionally we travelled a bit further afield to the Warrington Show, held in Walton Hall Gardens. Whenever we attended these events, Eric would inevitably spend some of the day walking around the horseboxes, seeking out familiar faces, catching up and reminiscing with old friends and making new ones. The old agricultural network was still alive and kicking.

It was not until 1963 that Eric finally found the time to enter one of these shows for himself. He entered Rupert in the tradesman's class at Liverpool Show. Unlike the other horses in the class, Rupert was still an active working horse and he put in a full morning shift delivering the milk before taking part in the show. He was not at all impressed when the heavy horses were paraded around the ring with all their ornate harness, ribbons and jingling bells – he showed his displeasure by rearing at them. The Rupert and Eric partnership came a creditable fourth in the class.

G. R. SUMNER, M.R.C.V.S.
VETERINARY SURGEON
58 HIGH STREET
WOOLTON, LIVERPOOL 25
L25 7TF

Telephone: 051 - 428 1024

SURGERIES: TUESDAYS and THURSDAYS 6 - 7 p.m.
SATURDAY 12 - 1 p.m.
Monday, Wednesday, Friday — Ring for appointment

Business card –– Sumner's Veterinary Surgery, Woolton Village. (Author)

Eric with Rupert at Liverpool Show, tradesman's class, 1963. (Author)

Eric with Rupert at Liverpool Show, tradesman's class, 1963. (Author)

1963 also marked the centenary of Orlando and George Joy opening a branch of the family milk business in Garston. The local paper ran the following story:

The local paper ran the following story under the headline, 'The Horse Makes Every Day A "JOY" Day, 1963. (Author)

Through the streets of the city at rush hour every day travels a monument to Garston's history ... a horse and trap.

For exactly 100 years the horses of the Joy family have roamed the streets of Liverpool on their daily milk rounds. Through the age of steam, petrol and electricity, horses have always been the first priority with the Joy family whose milk round goes back to the days before Garston Docks existed. It all started back in 1863 when granduncle Orlando Joy first owned a farm down by the river where the Stalbridge Docks now stand.

The farm – Dale farm – was just the beginning of things for the Joys, but when they were forced to move because of the construction of the dock it saw the start of the battle between the Joys and the industrial revolution.

Next move was made to some land on Horrocks Avenue where the farm stayed for about 70 years. But the Joys were fighting a losing battle, for in 1951 houses were built on Horrocks Avenue. However, the long-standing reputation of the Joys moved with them to their present stables in Duke Street.

Now all their milk comes from local firms, but the tradition of the Joys still holds strong and will for some time. In the words of Mr A Joy: 'Keeping horses is an art, once we got vans the art would disappear. We will always have horses.'

The three horses, Danny (aged 22), Rupert and Peggy, are famed throughout the district. Only the other week Rupert won fourth place at the Liverpool Show in the trade class.

Danny – 'the old age pensioner' of the stables – raised £10 for Liverpool's Handicapped Children at a fete recently when, with the trap, he provided the children with sixpenny rides.

Asked if he would ever go over to electric milk floats Mr Joy said: 'We've had horses for 100 years, why should we change now?'

There was one other family tradition into which Eric inducted his three young children – taking holidays in Yorkshire. The destination for our first holiday in the Dales was the cradle of the Joy family – the village of Hebden. We stayed in a self-catering chalet at the holiday centre run by Wharfedale Holidays. The centre was originally built in 1909 by the Co-operative Holidays' Association and reflected Wharfedale's emergence as a destination for tourists (David Joy, 2002). We spent many enjoyable days there exploring the local countryside and searching for clues as to our ancestry.

Despite stubbornly resisting the forces of change, time was inevitably catching up with Wellington Dairy. By this time the mark-up on each pint of milk was minimal and the business's profitability was decreasing. Plus, it wasn't just Danny who was 'the old age pensioner'. By the late 1960s both Percy and George were septuagenarians. Percy was already in a state of semi-retirement as he no longer went out on the rounds, and now it was finally George's turn. It was clear that Eric would be unable to continue running the dairy on his own, so the sad decision was made to sell the family business.

The hundred-year history of the Joy family dairying in Garston reflects the social and local history of the time. Daniel Joy and his siblings were part of the mass exodus from the countryside to the city, which took place during the mid-1800s. Although he was at one time farming at Dale Farm, that land was needed for the expansion of Garston Docks and its associated railway infrastructure. Daniel's son, Anthony Joy, was farming land off Island Road, but had to give up that land at the turn of the century when the need for public open space resulted in the creation of Garston Park. It was the expansion of the town of Garston

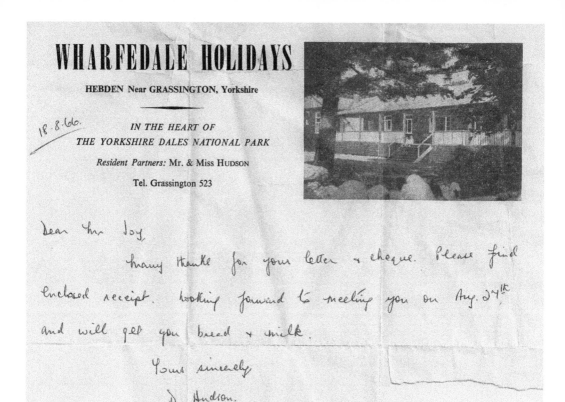

WHARFEDALE HOLIDAYS

HEBDEN Near GRASSINGTON, Yorkshire

18.8.66.

*IN THE HEART OF
THE YORKSHIRE DALES NATIONAL PARK*

Resident Partners: Mr. & Miss HUDSON

Tel. Grassington 523

Dear Mr Joy,

Many thanks for your letter & cheque. Please find enclosed receipt. Looking forward to meeting you on Aug. 27th, and will get you bread + milk.

Yours sincerely

D. Hudson.

Confirmation of holiday – Wharfedale Holidays, 18 August 1966. (Author)

and the need for more housing that led to the Joy family having to leave the fields off The Avenue; and then, the post-war baby boom, with its consequent need for bigger and better schools, resulted in the Joys giving up the field on Horrocks Avenue. Without land, the family were no longer farmers; without cows, they were no longer cowkeepers. Without a dairy, they were no longer dairymen.

The story of the end of Wellington Dairy is told in Dave Joy's book *My Family and Other Scousers* (2014). Percy and Ellen continued to live at 37 Wellington Street, with George and his wife living next door. But, the yard, shippon and stables were sold and good homes were found for the horses. The two milk rounds were sold as going concerns to a milkman who delivered the daily pint using an electric milk float. For the Joy family, a way of life that had spanned over a century had finally come to an end.

Although the Joys of Wellington Dairy were one of the last remaining examples of a Dales family who had become Liverpool cowkeepers and dairymen, they were not *the* last. The Liverpool and District Cowkeepers Association continued up until 25 April 1975, when it was finally wound up by the secretary, Mr T. G. Hogg (Mellor, 1978). It was just four months later that Mr Joe Capstick, of 4 Marlborough Road, moved his cows out of the city – to Brantbeck Farm, near Lancaster – and the last Liverpool cowkeeper was gone.

Above: Eric on holiday in Wharfedale, *c.* 1966 (Author)

Right: Percy and Eric Joy – last cowkeepers in the family line, *c.* 1966. (Author)

A Dairyman in Retirement

Eric's preparations for the closure of the dairy included securing alternative employment. He did this with the assistance of his brother-in-law, Edwin, who put in a good word for him down at Garston Docks. Eric began work for Irish Sea Ferries as a 'checker', which meant managing the inventory of all the recently introduced steel containers coming in and out of the docks. He found that he already knew many of the people he would be working with, as most came from Garston and some were his milk customers.

Eric retired from the docks in 1984. Although he had enjoyed his time working as part of the dock community, he was eager to retire in order to return to his first love – working with horses. Plus, by then he had three grandchildren to dote upon.

He kept himself busy in many ways. Locally, there was the Beechley Stables riding school for the disabled, located on Harthill Road. At the time this facility was being run by Liverpool City Council, but it was also very much dependent upon a dedicated group of volunteers. As well as helping out with the care of the horses, Eric undertook a project of his own design – constructing a cart that would accommodate a wheelchair in order for young people with mobility impairment to be able to learn how to drive a horse. Another locally based interest that earned him a bob or two was working for Olive Mason, driving a brougham for weddings and also for the annual Liverpool Lord Mayor's Parade.

Eric reignited his Veterinary Corp. network and travelled around the country, visiting friends. He attended agricultural shows and horse fairs and was invited onto the judging circuit; he became a regular steward at the Woolton Show and at the Royal Lancashire Agricultural Show. And then he became a television star! He worked for Granada Studios as a hansom cab driver when they were shooting the Sherlock Holmes series, starring Jeremy Brett. He would go to wardrobe and be kitted out with Victorian garb including top hat, and then to makeup, where they would give him long fuzzy cheek whiskers. Indoor filming took place at the studios in Manchester, while for outdoor work they went on location to Tatton Park in Cheshire.

It was wonderful for Eric to be back in his element, working with horses. He had a very happy retirement, especially as he was able to share a lot of what he was doing with his grandchildren, Victoria, Anthony and Heather. Inevitably, by the time he was in his eighties he had to cut back on some of his gallivanting around the country.

A dairyman in retirement – Beechley Stables Riding School, Allerton, Liverpool. (Author)

A dairyman in retirement – driving a wedding brougham. (Author)

A dairyman in retirement – two in hand. (Author)

A dairyman in retirement – teaching the next generation. (Author)

A dairyman in retirement – Liverpool Lord Mayor's Parade. (Author)

A dairyman in retirement – Judging at the Royal Lancashire Show. (Author)

A dairyman in retirement – stewarding at the Royal Lancashire Show. (Author)

Eric with actor Jeremy Brett (Sherlock Holmes), *c.* 1985. (Author)

It was on 28 February 2007, at the age of eighty-seven, that Eric Joy passed away quite suddenly. His funeral took place the following week and it was fitting that his coffin was borne by a horse-drawn hearse. He took one last trip down St Mary's Road, through Garston village, before being interred in the Joy family grave at St Michael's Church, Garston.

The horse-drawn hearse was a fitting mode of transport for Eric's final journey, 2007. (Author)

References

Atkins, P. J., 'The Pasteurization of England: the science, culture and health implications of food processing, 1900–1950' (University of Durham: accessed on www.academia.edu on 27 April 2015).

Atkins P. J., *Liquid Materialities: A History of Milk, Science and the Law* (2010).

Callaghan J., *Candles, Carts & Carbolic: A Liverpool Childhood Between the Wars* (2011).

Daine, H. S., 'A Short Account of Lancashire Farming in 1899', *Journal of the Royal Lancashire Agricultural Society* (1900).

Enock, A. G., *This Milk Business – A Study from 1895 to 1943* (1943).

Finegan F. W., *A Bit Akin – The Story of a North Craven Farming Family* (1994).

Grundy J., *Origins of Liverpool Cowkeepers* (Unpublished thesis, 1982).

Grundy J., 'Purpose-Built Premises for Town Cowkeepers in Liverpool', *Journal of the Historic Farms Building Group*, 4 (1990).

Harris A., 'The Milk Supply of East Yorkshire, 1850–1950', East Yorkshire Local History Society. Local History Series, 33 (1977).

Hartley M.and J. Ingilby, *Life and Tradition in the Yorkshire Dales* (1997).

Hill. H., 'Liverpool – Last Stronghold of the Town Cowkeepers', *Dairy Engineering* (1956).

Holmes J., 'The Cowkeepers of Liverpool'. *Sedbergh Historian*, Vol. IV, No. 5 (2002).

Hope E. W., *Health at the Gateway – Problems and International Obligations of a Seaport City* (1931).

Jenkins A., *Drink a Pinta – The Story of Milk & the Industry that Serves It* (1970).

Joy, Dave, *My Family and Other Scousers – A Liverpool Boy's Summer of Adventure in '69* (2014).

Joy David, *Uphill to Paradise – The Story of Hole Bottom Hamlet and Jerry & Ben's* (1991).

Joy, David, *Hebden – The History of a Dales Township* (2002).

Mackenzie, K. J. J., 'The Dual Purpose Cow at Liverpool', *Journal of The Royal Agricultural Society of England*, 71. (1910).

Marriott G. (Ed.), *Those Who Left the Dales* (2010).

Marshall, J., *The Lancashire and Yorkshire Railway Vol. 3* (1972).

Mellor P. J., 'Cow-keepers from the Pennine Dales', *The Dalesman* (1978).

Price B., 'Garston Farms and Past Inhabitants', *Past & Present*, 2 (2000).

Phelps T., *The British Milkman* (2011).

Scobie J., 'Cowkeepers from the Dales', Sedbergh Historian, Vol. V, No. 5 (2008).

Stoner F., 'Management and Working of the Liverpool Town Dairies', *Journal of the Royal Manchester, Liverpool and North Lancashire Agricultural Society* (1883).

Sumner J., *The Story of the Liverpool Cowkeeper* [Copy of an essay by John Sumner, MAFF Dairy Husbandry Adviser for Lancashire, held by Sedbergh & District History Society] (1970).

Swift J. M., *The Story of Garston and its Church* (1937).

Tansley, A. G., *Britain's Green Mantle* (1949).

Steere-Williams, J., 'Milking Science for its Worth: The Reform of the British Milk Trade in the Late Nineteenth Century', *Agricultural History*, Vol. 89, No. 2. (2015).

Yorkshire Vernacular Buildings Study Group, *Far Rams Close Building Report* (1977).

Acknowledgements

All photographs and images are from the family of the author's private collection with the exception of the following: Plans of Far Rams Close (Courtesy of the Yorkshire Vernacular Buildings Study Group), Traditional Milking at Dugdale's Farm, Grassendale Road, Garston (1950) (Courtesy of Frank Smallpage), Atkinson Dairymen, Heathfield Road, Wavertree (1915) (Courtesy of Mike Chitty, The Wavertree Society), 'There's A Bull Loose!' Cattle being herded along Lime Street (1931) (Reproduced under licence from British Pathe), Greenwood's Dairy, High Street, Wavertree (1887) (Courtesy of Mike Chitty, The Wavertree Society), Royal Lancashire Agricultural Show 1905 – Showground Plan (Courtesy of Mike Chitty, The Wavertree Society), The Royal Agricultural Show, Wavertree Playground (1910) (Courtesy of Mike Chitty, The Wavertree Society), Church Road and Railway Street, Garston (Courtesy of Garston & District Historical Society), View from St Michael's Church tower showing Railway Street, Garston (Courtesy of Garston & District Historical Society), Cows being herded along Aigburth Road, Garston (1950) (Courtesy of Frank Smallpage), Bryan Blackwell – Cowkeeper (Courtesy of Andrew Mellin), John Carr Davidson – Cowkeeper (Courtesy of Carole and Clive Davidson), J. C. Davidson – Business card (Courtesy of Barbara Price), Cows at Garston Cross – Corner of Chapel Road and Granville Road (Courtesy of Garston & District Historical Society), Home on leave. Eric, best man at Wilfred Percival's wedding – with bridesmaid, Joyce Bridge (1943) (Courtesy of Joyce Johnson), Liverpool & District Cowkeepers Association Annual Dinner (1952) (Courtesy of Gilbert and Margaret Rowlands), Liverpool Cowkeepers Annual Dinner. Exchange Hotel, Liverpool (1954) (Courtesy of Barbara Baker), 'A pride in doing a good job is general among the City's cowkeepers'. [*Photograph of Ken Wainwright*] (Courtesy of Frank Smallpage).

Special thanks to the following for their anecdotal contributions: Ron Atherton, Barbara Baker, Carole and Clive Davidson, Joyce Johnson, Arthur Jones, Christine Joy, Barry Joy, John Moor, Tom Ockleshaw, Barbara Price and Alan Wilkinson. Also, to: Trivena McCarlie at mctriv.com, for her artwork in re-creating Far Rams Close, and to Ronnie Hughes for hosting a cowkeeping forum on his blog site 'A Sense of Place'.

Grateful thanks to the following for their helpful communications during my research: Barbara Price (Garston & District Historical Society); David Cant (Yorkshire Vernacular Buildings Study Group); Mike Chitty (The Wavertree Society); Elspeth Griffiths (Sedbergh & District History Society); David Joy MBE (fourth cousin once removed); Alan Passmore; Jean Robinson (Friends of Raikes Road Burial Ground); Duncan Scott; Ian Sime, and Amy Smith (Country Publications).